PROJECT AND COST ENGINEERS' HANDBOOK

Jim SHAH

Edited by F. C. Jelen, Ph.D.

Contributors

D. L. Anderson
J. W. Higginbotham
E. E. Stackhouse
R. L. Dodds
F. C. Jelen
W. Stelly
J. L. Hamlin
J. J. O'Driscoll
A. J. Wallace

ISBN: 0-930284-06-2

The American Association of Cost Engineers
308 Monongahela Building
Morgantown, West Virginia 26505

Jim Shah

Preface

"Project and Cost Engineers' Handbook" came into being as a direct result of the A.A.C.E. Certification program. As the 'grandfathering' phase ended and requirements for Certification included an examination, it became apparent that most of us needed considerable study if we were to become recognized by our peers as "well rounded and competent in all phases of Cost Engineering."

An all day short-course as well as a study guide with example questions had been developed by the Certification Board. However, results from the first examinations soon proved that most of us tend to 'specialize' and quickly loose our sharpness in many of the other important phases of Cost Engineering.

The Gulf Coast Section siezed upon this opportunity and developed a 14 week study course in which each phase of Cost Engineering was developed in detail by a Cost Engineer especially skilled in each phase of Cost Engineering.

Because of the success of this study course, the Gulf Coast Section petitioned the A.A.C.E. Board for help and permission as well as sanction to condense this study course into the present book.

All of us are especially indebted to Aaron Cohen for his personal efforts and persistency. Also, without the dedication and work of Dr. Fred C. Jelen of Lamar University, this book would not have been published.

As Cost Engineers, all of us can use this book--and learn from using it. We are appreciative of those who gave their time and efforts in preparing and presenting the basic material upon which this book is drawn.

Also, acknowledgement is given to those who have granted permission to use copywrited materials within the text.

This is just one more example of the members of A.A.C.E. willingness to share the skills and knowledge of Cost Engineering.

Don't walk in front of me
I may not follow.
Don't walk behind me
I may not lead.
Just walk beside me
Andbe my friend.

Thomas C. Ponder, Fellow, A.A.C.E.
P.E., C.C.E.
Houston, Texas 27 August 1979

American Association of Cost Engineers

Review Committee

Lawrence J. Bloch, C.C.E.

Gordon F. Gantz, C.C.E.

J. A. Foushi, C.C.E.

Coordinating Committee

Aaron J. Cohen, C.C.E.

Thomas C. Ponder, C.C.E.

Editor

Dr. F. C. Jelen, C.C.E.

List of Contributors

Anderson, Andy, B.S., Manager, Cost Engineering Department, Brown & Root, Inc., Houston, Texas. Mr. Anderson received his B.S. from Texas A & M University in 1956. He began his career in construction as a structural engineer working in St. Louis, Missouri. He joined Brown & Root as an estimator in 1961. In an effort to gather cost information for the Estimating Department of Brown & Root, he was sent to construction projects throughout the United States and Canada. The importance of the cost engineering function prompted the company to establish a separate Cost Engineering Department of which he became Assistant Manager in 1967 and Manager in 1970. Mr. Anderson has been active in the A.A.C.E. since he became involved in Cost Engineering. He served as Secretary in 1969-70, President in 1970-71 and as Director 1971-72 of the Gulf Coast Section.

Dobbs, R. L., Independent Consultant, Houston, Texas. Mr. Dobbs attended Lehigh University 1946-48 studying chemical engineering; the University of Texas 1954-56 studying management; and MBA studies at the University of Houston. He has spent 21 years in the process industry; 10 years managing/performing planning, scheduling, cost engineering and progress reporting for grass roots refineries, major refinery expansions, petrochemical units, gas plants, offsite facilities, pipelines and marine facilities. Eleven years in operations with petrochemical and food processing firms. Presently he is an Independent Consultant. He is a Certified Cost Engineer and served as Vice-President and President of the A.A.C.E. Gulf Cost Section.

Hamlin, Jerry, M.B.A., Senior Consultant, Productivity Information systems, Technical Systems Division, Cities Service

Co., Tulsa, Okla. He is the 1979-80 President of the Northeast Oklahoma Section and also a member of the A.A.C.E. Productivity Committee. Jerry received the 1979 Award of Excellence in Publications in behalf of his efforts to initiate the Cost Engineering skills course leading to the publication of this book. Mr. Hamlin is an active member and Director of the American Institute of Industrial Engineers. He received a B.S. in industrial engineering from Oklahoma State University, a M.S. in industrial engineering from University of Tennessee, and an M.B.A. from the University of Tulsa. He is a Registered Professional Engineer and a Certified Cost Engineer.

Higginbotham, J. W., Cost Engineer, St. Regis Corporate Engineering, Houston, Texas. Mr. Higginbotham received diplomas from the International Correspondence School in drafting, design and engineering courses; LaSalle Extension University in Accounting in 1965, and Hotel, Motel Management in 1972. He received the grade of Senior Engineering Technician in 1966 and also holds a Real Estate License for the State of Texas. He is a Member of A.A.C.E. and served as Treasurer and two years on the Board of Directors of the Gulf Coast Section. He is a member of the National American Society of Professional Estimators and the Houston Chapter. He is an Associate Member of the Gulf Coast Board of Realtors, Associate Member of the Texas Association of Realtors, Member of the American Society of Engineering Technicians, and is a Certified Cost Engineer. He was an Instructor for Profitability & Planning Section of Cost Engineering Course at the University of Houston as well as an Instructor for the Cost Engineering Skills Course sponsored by the Gulf Section of A.A.C.E. on the same subject.

Jelen, F. C., Ph.D., Professor of Chemical Engineering, Lamar University, Beaumont, Texas. Dr. Jelen holds S.B. and S.M. degrees from the Massachusetts Institute of Technology and A.M. and Ph.D. degrees from the Harvard Graduate School of Arts and Sciences. He spent 25 years in industrial development and research and then in 1961 assumed his present position as Professor of Chemical Engineering at Lamar University. He is a member of the A.I.Ch.E., the A.A.C.E. He is a licensed professional engineer in Texas and New York and is listed in Who's Who In Engineering, Who's Who in Consulting, and American Men of Science. He has served as a consultant to the American Economic Foundation and has published many articles on engineering economics. He has pioneered in promoting continuing education on this subject for the A.I.Ch.E. and has lectured in Mexico and Hawaii. He is a popular lecturer, having received an award from A.I.Ch.E. for best presentation of a paper and has served on the speakers' bureau of the A.I.Ch.E. He was the editor of Cost and Optimization Engineering, the first book produced by A.A.C.E. He has recieved the Award of Merit and Life Membership from the A.A.C.E.

O''Driscoll, J. J., M.S., Staff Engineer of Cost Engineering Department, Shell Oil Co., Houston, Texas. He received his B.S. and M.S. in civil engineering from Carnegie Mellon University. He is a Fellow of A.A.C.E., recipient of the Award of Merit, served as Chairman of preparations and publications, A.A.C.E. Transactions, 1970-72, A.A.C.E. Treasurer, 1972-78. In the Gulf Coast Section he has served as Director, Secretary, and presently President. He has also conducted classes in Cost Engineering at the University of Houston.

Stackhouse, E. E., M.A., Project Control Supervision, Petro-Marine Engineering, Inc., Houston, Texas. He has twelve years experience in plant and project engineering, project management and control, and administration for the petroleum and petrochemical industries. He is active in the A.A.C.E., Gulf Coast Section and is past Secretary and vice president. He was a lecturer for the 1978 Project/Cost Engineers' Skills Course and the 1979 Spring Workshop sponsored by the Gulf Coast Section of A.A.C.E. Mr. Stackhouse received a B.S. in mechanical engineering from Louisiana Tech University and performed graduate sudies at Georgia Tech. He is a member of A.A.C.E., Pi Tau Signa, Project Management Institute and Tau Beta Pi.

Stelly, Wil, Manager of Cost Engineering, C-E Lummus, Houston, Texas. He is a graduate industrial engineer of International Correspondence Schools with additional studies in mechanical engineering at Lamar University and management training with Rensselaer Polytechnic Institute. He is a senior member of A.I.I.E. and a Certified Cost Engineer. With more than 37 years in construction, estimating and cost control work, Mr. Stelly is one of the industry's leading authorities on cost engineering.

Wallace, A. J., B.B.A., Senior Project Controller, C-E Lummus, Houston, Texas. He is a graduate of the University of Houston with over 20 years experience in financial accounting.

CONTENTS

Part 2. Classification of cost

Part 3. Simulation

1. COST ENGINEERING WORKSHOP

F. C. Jelen
Lamar University, Beaumont, Texas

This chapter of the course is generally given by itself as a one to one and one-half day course for an intensive review for the certified cost engineering examination. The purpose is to help most in those topics where the participants will require guidance. Hence the allotment of space for topics may seem uneven, but is prompted by the performance of those taking the previous examinations.

The examination is comprised of four equal parts each being a 1.5 hour written examination. Two are given in the morning and two in the afternoon of the same day. The four parts are

- Part 1. Basic general knowledge
- Part 2. Costs and cost control
- Part 3. Associated cost topics
- Part 4. Management aspects

The suggested study guide for these four parts is given on the succeeding pages. The examination is prepared from the suggested study guide. Parts 1 and 4 have been open book and Parts 2 and 3 have been closed book.

The book "Cost and Optimization Engineering" sponsored by the AACE and edited by F. C. Jelen is perhaps the best general text for the examination. Parts 2 and 3 in particular, however, will require supplementary knowledge and a list of references is given at the end of this chapter.

COST COMPARISONS. The mathematics of cost comparisons involves the following

- Time value of money
- Relationships among the kinds of cost
- Methods for handling unequal durations

An important principle is that amounts occurring at different instants of time cannot be added directly, but first must be reduced to some common instant of time before adding. Conversion of an amount from one instant to another instant is made by the time value conversion relationship

$$A_2 = A_1(1 + i)^{\pm n}$$

Where A_1 is the amount at time 1, A_2 is the equivalent amount at time 2, i is the time value of money expressed as a decimal for one period (usually a year) and n is the number of periods of time (years) between A_1 and A_2. If moving from A_1 to A_2 is moving with the calendar then n is positive and A_2 is greater than A_1. When moving against the calendar n is negative and A_2 is less than A_1.

SUGGESTED STUDY GUIDE

Certified Cost Engineer Examination
American Association of Cost Engineers

Part 1. Basic general knowledge

1. Equivalence and cost comparisons
   a. Time value of money
   b. Discrete interest
   c. Continuous interest
   d. Detailed comparison of alternatives
   e. Replacement and displacement problems
   f. Life cycle costing
   g. Benefit to cost ratio

2. Taxes
   a. Depreciation
   b. Depletion

3. Inflation

4. Learning curve

5. Profitability- venture analysis
   a. Return on investment
   b. Payout time
   c. Present value
   d. Discounted cash flow rate of return

Part 2. Costs and cost control

1. Construction costs

2. Manufacturing costs

3. Capital cost estimation

4. Operating cost estimation

5. Individual elements of cost
   a. Labor
   b. Materials
   c. Utilities
   d. Transportation

6. Costing plants and projects

7. Costing products and services

8. Cost control
   a. Planning
   b. Scheduling
   c. Management of
   d. CPM and other network techniques

Part 3. Associated cost topics

1. Cost accounting
   a. Direct costs
   b. Indirect costs
   c. Standard costs
   d. Joint costs, transfer pricing etc.
   e. Overhead
2. Design considerations and cost
3. Economics
   a. Market factors
   b. Current costs
   c. Plant site selection
4. Forecasting
5. Productivity
6. Cost indexes
7. Value analysis
   a. Appraisals
   b. Pricing
   c. Quality considerations
8. Foreign countries
9. Presentation of requests and reports

Part 4. Management aspects

1. General
   a. Project management
   b. Project staffing
   c. Personnel administration--human relations
   d. Payroll
   e. Industry knowledge
   f. Information sources
2. Contractural elements
   a. Bidding
   b. Contracting methods
   c. Legal considerations
3. Budgeting
4. Society
   a. Role of government
   b. Social obligations
   c. Ethics
5. Use of computers
6. Management science
   a. Optimization techniques
   b. Inventory problems
   c. Queuing problems
   d. Linear programming
   e. Dynamic programming
   f. Geometric programming
   g. Monte Carlo technique
   h. Break-even analysis
   i. Uncertainty

There are four principal timings for costs.

- P present value, a single amount
- S future amount, a single amount
- R a uniform end-of-year amount
- R_b a uniform beginning-of-year amount

The relationships between those amounts can be expressed as a factor F with subscripts to express the conversion. Thus

- $F_{RP,i,n}$ is the factor which converts $1 of R to a present worth at a decimal rate i per period for n periods
- $F_{SR,i,n}$ is the factor that converts $1 of future amount S n periods from now to a uniform end-of-period amount R for n periods at a decimal rate i per period.

A summary of the relationships is given in Table 1-1 on the following page.

A second important principle is that when making cost comparisons involving different durations, the comparison must be made on equivalent costs for a common duration. There are two methods for obtaining a common denominator for unequal durations

1. Unacost - an equivalent end-of-year annual cost
2. Capitalized cost - an equivalent present cost for infinite service.

Unacost is denoted by R and capitalized cost by K. They are related since

$$R = iK$$

Both unacost and capitalized cost are merely mathematical conveniences for obtaining common denominators of service lives. There is no assumption that a one year life or a perpetual life is required.

Table 1-1

Summary of S, P, R, and R_b Relationships

Item	*Conversion*	*Algebraic relationship*	*Relationship by factor*	*Name of factor*
1	*P* to *S*	$S = P[(1+i)^n]$	$S = PF_{PS,i,n}$	Compound-interest factor
2	*S* to *P*	$P = S[(1+i)^{-n}]$	$P = SF_{SP,i,n}$	Present-worth factor
3	*R* to *P*	$P = R\dfrac{(1+i)^n - 1}{i(1+i)^n}$	$P = RF_{RP,i,n}$	Unacost present-worth factor
4	*P* to *R*	$R = P\dfrac{i(1+i)^n}{(1+i)^n - 1}$	$R = PF_{PR,i,n}$	Capital-recovery factor
5	*R* to *S*	$S = R\dfrac{(1+i)^n - 1}{i}$	$S = RF_{RS,i,n}$ $S = RF_{RP,i,n}F_{PS,i,n}$	Equal-payment-series future-worth factor
6	*S* to *R*	$R = S\dfrac{i}{(1+i)^n - 1}$	$R = SF_{SR,i,n}$ $R = SF_{SP,i,n}F_{PR,i,n}$	Sinking-fund factor
7	R_b to *P*	$P = R_b(1+i)\dfrac{(1+i)^n - 1}{i(1+i)^n}$	$P = R_b(1+i)F_{RP,i,n}$	
8	*P* to R_b	$R_b = \dfrac{P}{1+i}\dfrac{i(1+i)^n}{(1+i)^n - 1}$	$R_b = \dfrac{P}{1+i}F_{PR,i,n}$	
9	R_b to *S*	$S = R_b(1+i)\dfrac{(1+i)^n - 1}{i}$	$S = R_b(1+i)F_{RS,i,n}$ $S = R_b(1+i)F_{RP,i,n}F_{PS,i,n}$	
10	*S* to R_b	$R_b = \dfrac{S}{1+i}\dfrac{i}{(1+i)^n - 1}$	$R_b = S\dfrac{1}{1+i}F_{SR,i,n}$ $R_b = S\dfrac{1}{1+i}F_{SP,i,n}F_{PR,i,n}$	

Jelen, F. C. Cost and Optimization Engineering
McGraw-Hill Book Co. (1970)

Reprinted with permission

A summary for Unacost and capitalized cost for various items of cost is given in Table 1-2.

TAXES. In practice cost comparisons are made on an after tax basis. Depreciation must be considered in determining the tax base on which taxes are paid. Depreciation arises when an expenditure outlasts one year and a depreciation schedule is a means of allocating the expenditure as an expense over the years. The accountant's application of depreciation for each year is a Depreciation Expense account and it is an expense for tax purposes. There are four principal kinds of depreciation schedules.

1. Straight line depreciation (SL). The depreciation is constant for all the years and is for each year

$$\frac{C_d}{n}$$

where C_d is the depreciable first cost and n is the depreciable life in years.

2. Declining balance depreciation(DB). A factor F_{DB} is determined from the salvage value and the life and used as follows to obtain a decreasing depreciation for the individual years.

$$F_{DB} = 1 - \sqrt[n]{\frac{C_{SAL}}{C_i}}$$

C_i is the initial cost. The depreciation for the years is obtained from the following sequence

$$D_1 = C_i F_{DB}$$

$$D_2 = C_i (1 - F_{DB}) F_{DB}$$

$$D_3 = C_i (1 - F_{DB})^2 F_{DB} \text{ etc.}$$

Table 1-2

Summary of Unacost and Capitalized-cost Relationships—No Tax

Item no.	Item	Present worth as a cost for n years duration	Unacost: Algebraic expression	Unacost: Factor expression	Capitalized cost: Algebraic expression	Capitalized cost: Factor expression
1	C_i Initial cost for n-year life	C_i	$C_i \frac{i(1+i)^n}{(1+i)^n - 1}$	$C_i F_{PR,i,n}$	Multiply corresponding algebraic expression for unacost by $\frac{1}{i}$	$C_i F_{PK,i,n}$
2	R Uniform end-of-year annual cost	$R \frac{(1+i)^n - 1}{i(1+i)^n}$	R	$R1$ (factor is unity)		$R \frac{1}{i}$
3	R_b Uniform beginning-of-year annual cost	$R_b(1+i) \frac{(1+i)^n - 1}{i(1+i)^n}$	$R_b(1+i)$	$R_b(1+i)$		$R_b F_{PK,i,1}$
4	C_{ex} Irregular cost at end of xth year for article lasting n years	$C_{ex} \frac{1}{(1+i)^x}$	$C_{ex} \frac{1}{(1+i)^x} \times \frac{i(1+i)^n}{(1+i)^n - 1}$	$C_{ex} \frac{1}{(1+i)^x} F_{PR,i,n}$		$C_{ex} \frac{1}{(1+i)^x} F_{PK,i,n}$
5	C_{bx} Irregular cost at beginning of xth year for article lasting n years	$C_{bx} \frac{1}{(1+i)^{x-1}}$	$C_{bx} \frac{1}{(1+i)^{x-1}} \times \frac{i(1+i)^n}{(1+i)^n - 1}$	$C_{bx} \frac{1}{(1+i)^{x-1}} F_{PR,i,n}$		$C_{bx} \frac{1}{(1+i)^{x-1}} F_{PK,i,n}$
6	C_{nd} Nondepreciable first cost, e.g., land or an article that lasts forever	$C_{nd} \frac{(1+i)^n - 1}{(1+i)^n}$	$C_{nd}\, i$	$C_{nd}\, i$		$C_{nd}\, 1$ (factor is unity)
7	C_{sal} Salvage value at end of nth year, a receipt	$-C_{sal} \frac{1}{(1+i)^n}$	$-C_{sal} \frac{i}{(1+i)^n - 1}$	$-C_{sal} \frac{1}{(1+i)^n} \times F_{PR,i,n}$		$-C_{sal}(F_{PK,i,n} - 1)$ or $-C_{sal} \frac{1}{(1+i)^n} F_{PK,i,n}$
8	Initial cost C_i and salvage value C_{sal} taken together	$(C_i - C_{sal}) + C_{sal} \times \frac{(1+i)^n - 1}{(1+i)^n}$	$(C_i - C_{sal}) \frac{i(1+i)^n}{(1+i)^n - 1} + iC_{sal}$	$(C_i - C_{sal}) \times F_{PR,i,n} + iC_{sal}$		$(C_i - C_{sal}) F_{PK,i,n} + C_{sal}$

Jelen, F. C. Cost and Optimization Engineering McGraw-Hill Book Co. (1970)

Reprinted with permission

3. Sum-of-the-years digits depreciation(SD). The method is best illustrated by an example, say for a 4 year life where sum-of-the-years digits is 4 + 3 + 2 + 1 = 10 (In general the sum is $n(n + 1)/2$). The depreciation by years for a depreciable cost C_d is

$$D_1 = C_d \frac{4}{10}$$

$$D_2 = C_d \frac{3}{10}$$

$$D_3 = C_d \frac{2}{10}$$

$$D_4 = C_d \frac{1}{10}$$

The yearly depreciation decreases with time.

4. Double declining balance depreciation(DDB). The method resembles the declining balance method but the factor F_{DDB} is obtained from

$$F_{DDB} = \frac{2}{n}$$

and the decreasing depreciation for the years is

$$D_1 = C_i F_{DDB}$$

$$D_2 = C_i (1 - F_{DDB}) F_{DDB}$$

$$D_3 = C_i (1 - F_{DDB})^2 F_{DDB}$$

Relationships for various depreciation schedules are given in Table 1-3.

Although the double declining balance method requires a factor 2/n, it is permissible to select any smaller factor, such as 1.5/n.

Table 1-3

Relationships for Depreciation, *n*-year Life*

Item no.	Method	Annual depreciation for *m*th year	Accumulated depreciation for *m* years	Present worth of *m* years of depreciation
1	Straight line (SL)	$\frac{C_i - C_{sal}}{n}$	$(C_i - C_{sal})\frac{m}{n}$	$\frac{C_i - C_{sal}}{n} F_{RP,i,m}$
2	Sinking fund (SF)	$(C_i - C_{sal})F_{SP,i,n}F_{PR,i,n}F_{PS,i,m-1}$	$(C_i - C_{sal})\frac{F_{PS,i,m} - 1}{F_{PS,i,n} - 1}$	$(C_i - C_{sal})\frac{m}{1+i} F_{SP,i,n}F_{PR,i,n}$
3	Declining balance (DB)	$C_i\left(\sqrt[n]{\frac{C_{sal}}{C_i}}\right)^{m-1}\left(1 - \sqrt[n]{\frac{C_{sal}}{C_i}}\right)$ or $C_i(1 - F_{DB})^{m-1}F_{DB}$	$C_i\left[1 - \left(\sqrt[n]{\frac{C_{sal}}{C_i}}\right)^m\right]$ or $C_i[1 - (1 - F_{DB})^m]$	$C_i\left(1 - \sqrt[n]{\frac{C_{sal}}{C_i}}\right)\frac{1 - \left(\frac{\sqrt[n]{C_{sal}/C_i}}{1+i}\right)^m}{1 + i - \sqrt[n]{\frac{C_{sal}}{C_i}}}$
4	Double declining balance (DDB)	$C_i\left(1 - \frac{2}{n}\right)^{m-1}\frac{2}{n}$	$C_i\left[1 - \left(1 - \frac{2}{n}\right)^m\right]$	$\frac{2C_i}{n}\frac{1 - \left(\frac{1 - 2/n}{1+i}\right)^m}{i + 2/n}$
5	Sum of digits (SD)	$(C_i - C_{sal})\frac{n - m + 1}{0.5n(n+1)}$	$(C_i - C_{sal})\frac{m}{n(n+1)}(2n + 1 - m)$	$\frac{2(C_i - C_{sal})}{n(n+1)i}$ $\times [n - F_{RP,i,m} - (n - m)F_{SP,i,m}]$.
6	Units of production (UP)	$(C_i - C_{sal})\frac{M_m}{M}$	$(C_i - C_{sal})\frac{\sum_1^m M_m}{M}$	$\frac{C_i - C_{sal}}{M}\sum_1^m \frac{M_m}{(1+i)^m}$

* Book value at end of *m*th year = C_i less accumulated depreciation for *m* years. Depreciation remaining for $n - m$ future years = book value at end of *m*th year less C_{sal}. For double-declining-balance depreciation, terminal value is not related to salvage value, but total write-off should be to salvage value.

Jelen, F. C. Cost and Optimization Engineering McGraw-Hill Book Co. (1970)

It is permissible to combine assets in a group and grouping is used widely to avoid an unnecessary burden of a profusing of accounts. For more details see the AACE Cost Engineers' Notebook. D - 7.8 1965.

The different depreciation methods offer different procedures for allocating the depreciation over the individual years. The total depreciation allowable for tax purpose is the same for all methods. In no case may the book value fall below the predetermined salvage value as this would amount to excessive depreciation. Straight line depreciation is a uniform method. The other methods mentioned are accelerated methods with the rate of depreciation highest at the start. The double declining method gives the highest present value for the future depreciation followed by the sum-of-the-years method, the declining balance method, and the straight line method. A sinking fund depreciation method, not discussed here, has an increasing rate of depreciation with time but is not used by industry although public utilities sometimes choose it.

Unacost and capitalized cost are used on an after tax basis for making cost comparisons involving different durations. A summary of the after tax relationships is given in Table 1-4. It is instructive to note how the tax factor enters the Table particularly if the tax situation is spread over a number of years. Thus for Item 1 for a depreciable first cost of $1 spread over 4 years depreciation the cost diagram is

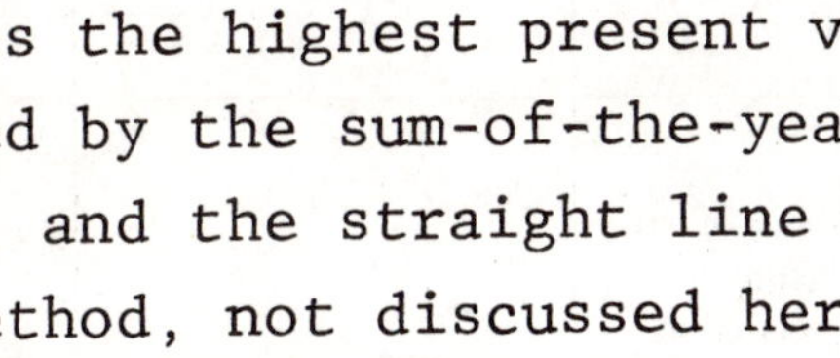

Where D is the fractional depreciation for a year and t the decimal tax rate. At each year end the tax base is reduced

Table 1-4

Summary of Unacost and Capitalized-cost Relationships with Tax

Item no.	Item	Present worth as a cost for n-year duration	Unacost: Algebraic expression
1	C_d Fully depreciable part of an initial cost for n-year life, n' years for tax purposes; $C_d = C_i - C_{sal}$	$C_d(1 - t\psi_{r_{n'}})$	$C_d(1 - t\psi_{r_{n'}}) \times \frac{r(1+r)^n}{(1+r)^n - 1}$
2	R Uniform end-of-year annual cost, depreciated fully at instant incurred	$R(1-t)\frac{(1+r)^n - 1}{r(1+r)}$	$R(1-t)$
3	R_b Uniform beginning-of-year annual cost, fully depreciated at end of year	$R_b\left(1 - \frac{t}{1+r}\right)(1+r) \times \frac{(1+r)^n - 1}{r(1+r)^n}$	$R_b\left(1 - \frac{t}{1+r}\right)(1+r)$
4	C_{ex} Irregular cost at end of xth year for article lasting n years, fully depreciated at instant incurred	$C_{ex}\frac{1-t}{(1+r)^x}$ or $\frac{C_{ex}}{(1+r)^x} \times [1 - t(1+r)\psi_{r_1}]$	$C_{ex}\frac{1-t}{(1+r)^x} \times \frac{r(1+r)^n}{(1+r)^n - 1}$
5	C_{bx} Irregular cost at beginning of xth year for article lasting n years, fully depreciated at end of year x	$C_{bx}\left(1 - \frac{t}{1+r}\right) \times \frac{1}{(1+r)^{x-1}}$ or $\frac{C_{bx}}{(1+r)^{x-1}}(1 - t\psi_{r_1})$	$C_{bx}\left(1 - \frac{t}{1+r}\right) \times \frac{1}{(1+r)^{x-1}} \times \frac{r(1+r)^n}{(1+r)^n - 1}$
6	C_{nd} Nondepreciable first cost such as land or an article that lasts forever	$C_{nd}\frac{(1+r)^n - 1}{(1+r)^n}$	$C_{nd}\,r$
7	C_{sal} Salvage value at end of nth year; treated as a nondepreciable first cost, an expense	$C_{sal}\frac{(1+r)^n - 1}{(1+r)^n}$	$C_{sal}\,r$

Jelen, F. C. Cost and Optimization and Engineering
McGraw-Hill Book Co. (1970)

Reprinted with permission

Table 1-4 continued

Unacost	*Capitalized cost*	
Factor expression for sum-of-digits depreciation	*Algebraic expression*	*Factor expression for sum-of-digits depreciation*
$C_d(1 - tF_{SDP,r,n'})F_{PR,r,n}$	Multiply corresponding algebraic expression for unacost by $\frac{1}{r}$	$C_d(1 - tF_{SDP,r,n'})F_{P\cdot,r,n}$
$R(1 - t)$		$R(1 - t)\frac{1}{r}$
$R_b(1 - tF_{SDP,r,1})(1 + r)$		$R_b(1 - tF_{SDP,r,1})F_{PK,r,1}$
$C_{ex}\frac{1}{(1+r)^x} \times [1 - t(1 + r)F_{SDP,r,1}]F_{PR,r,n}$		$C_{ex}\frac{1}{(1+r)^x} \times [1 - t(1 + r)F_{SDP,r,1}]F_{PK,r,n}$
$C_{bx}\frac{1 - tF_{SDP,r,1}}{(1+r)^{x-1}}F_{PR,r,n}$		$C_{bx}\frac{1 - tF_{SDP,r,1}}{(1+r)^{x-1}}F_{PK,r,n}$
$C_{nd}\,r$		C_{nd}
$C_{sal}\,r$		C_{sal}

Jelen, F. C. <u>Cost and Optimization Engineering</u>
McGraw-Hill Book Co. (1970)

Reprinted with permission

by D, the depreciation expense, and a tax benefit of Dt is saved. The present value of the above is

$$P = 1 - \Sigma \frac{D_n t}{(1 + r)^n} = 1 - t\psi$$

ψ is the present value of \$1 of future depreciation. It can be calculated for any depreciation method and for any n and r. The factor $F_{SLP,i,n}$ converts \$1 of future depreciation to a present value for straight line depreciation.

Example 1-1. Two machines have the following cost comparison. If money is worth 10% per year, the tax rate is 46%, and sum-of-year-digits depreciation is used, which machine is more economical? What is the savings per year?

	A	B
First cost, \$	18,000	24,000
Uniform end of year expense, \$	1,000	0
Salvage value, \$	500	0
Service life, years	2	3
Life for tax purpose, years	5	5

Factors are taken from the Appendix in "Cost and Optimization Engineering." For machine A refer to Table 1-4 and use the unacost expressions for items 1, 2, and 7 in order.

$C_d(1 - tF_{SDP,10\%,5})F_{PR,10\%,2}$

$= (18{,}000 - 500)[1 - .46(.80614)](.57619) \quad = \$6{,}344.2$

$R(1 - t) = 1{,}000(1 - .46) \quad = 540.0$

$C_{sal}(r) = 500(.10) \quad = 50.0$

$R_A = \$6{,}934.2$

For machine B by Item 1

$$C_d(1 - tF_{SDP,10\%,5})F_{PR,10\%,5}$$

$$= 24{,}000[1 - .46(.80614)](.40211) = R_B = \$6{,}071.9$$

Machine A is more economical with an end of each year savings after taxes of (6934.2 - 6071.9) = \$862.3. The yearly savings should be reported on a before tax basis and is 862.3/(1 - .46) = \$1,597.

The following example is an interesting application. The principles apply when taxes are included, but for simplicity the example omits taxes.

Example 1-2. A machine costs \$20,000 and lasts 6 years with \$1,000 salvage value at all times. Past records give the following data where costs are equivalent costs at the beginning of the year and include operating and repair costs.

Year	$
1	3,000
2	3,500
3	8,000
4	15,000
5	5,000
6	15,000

If money is worth 10% per year, how long should the machine be kept?

The solution is given in Table 1-5 in which is calculated the unacost for all possible durations of service.

Table 1-5
Solution to Example 1-2

Unacost basis 1 yr.

$$\left(20{,}000 + 3{,}000 - \frac{1{,}000}{1.10}\right) F_{PR,10\%,1} =$$

$$(23{,}000 - 909)(1.10) = 24{,}300$$

Unacost basis 2 yr.

$$\left(20{,}000 + 3{,}000 + \frac{3{,}500}{1.10} - \frac{1{,}000}{(1.10)^2}\right) F_{PR,10\%,2} =$$

$$(26{,}182 - 826)(0.57619) = 14{,}610$$

Unacost basis 3 yr.

$$\left(23{,}000 + \frac{3{,}500}{1.10} + \frac{8{,}000}{(1.10)^2} - \frac{1{,}000}{(1.10)^3}\right) F_{PR,10\%,3} =$$

$$(32{,}794 - 751)(0.40211) = 12{,}885$$

Unacost basis 4 yr.

$$\left(23{,}000 + \frac{3{,}500}{1.10} + \frac{8{,}000}{(1.10)^2} + \frac{15{,}000}{(1.10)^3} - \frac{1{,}000}{(1.10)^4}\right) F_{PR,10\%,4} =$$

$$(44{,}064 - 683)(0.31547) = 13{,}685$$

Unacost basis 5 yr.

$$\left(23{,}000 + \frac{3{,}500}{1.10} + \frac{8{,}000}{(1.10)^2} + \frac{15{,}000}{(1.10)^3} + \frac{5{,}000}{(1.10)^4} - \frac{1000}{(1.10)^5}\right) F_{PR,10\%,5} =$$

$$(47{,}479 - 621)(0.26380) = 12{,}361$$

Unacost basis 6 yr.

$$\left(23{,}000 + \frac{3{,}500}{1.10} + \frac{8{,}000}{(1.10)^2} + \frac{15{,}000}{(1.10)^3} + \frac{5{,}000}{(1.10)^4} + \frac{15{,}000}{(1.10)^5} - \frac{1000}{(1.10)^6}\right) F_{PR,10\%,6}$$

$$= (56{,}796 - 564)(0.22961) = 12{,}911$$

Minimum unacost by keeping machine 5 yr.

Jelen, F.C. Solutions Manual for Cost and Optimization Engineering
McGraw-Hill Book Co. (1970)
Reprinted with permission.

CONTINUOUS INTEREST. For continuous interest $e^{\pm in}$ is the operator that replaces $(1 + i)^{\pm n}$ for discreet interest. Thus

$$A_2 = A\ e^{\pm in}$$

where n is positive moving with the calendar, and negative when against the calendar.

Cost comparisons involving different durations are based on unaflow, $\bar{R}$, a uniform yearly annual flow. A summary of relationships for continuous interest is given in Table 1-6.

For more details on continuous interest see Chapter 4 of "Cost and Optimization Engineering".

INFLATION. Inflation is a factor in cost comparisons and can be accommodated for any decimal inflation rate d. The mathematical relationships based on capitalized cost for the various items of cost are given in Table 1-7. The Table should only be used for rate of return after taxes, r, greater than the inflation rate, d, the only practical consideration.

Instead of capitalized cost for a common denominator of all service lives, comparison involving different service lives can be compared on the bases of unaburden, R_D, which is an equivalent end of year burden (not dollars), but in dollars is an end of year cost increasing by the factor (1 + d) in actual dollars. Unaburden is readily obtained from the Table (for r > d) by multiplying the capitalized cost by (r - d), that is

$$R_D = (r - d)K$$

Example 1-3. A machine lasts 2 years with negligible salvage value, can be written off in 1 year for tax purposes, and

Table 1-6

Summary of Relationships for Continuous Interest

Item no.	*Item*	*Description*	*Algebraic relationship*	*Factor relationship*
1	P to S	Moves a fixed sum P to another instant of time n years with calendar	$S = Pe^{\bar{i}n}$	$S = PF_{PS,\bar{i},n}$ (App. 2, Table 1)
2	S to P	Moves a fixed sum S to another instant of time n years against calendar	$P = Se^{-\bar{i}n}$	$P = SF_{SP,\bar{i},n}$ (App. 2, Table 2)
3	$\bar{R}$ to P	Converts a unaflow $\bar{R}$ for n years to present worth at start of flow	$P = n\bar{R}\dfrac{1 - e^{-\bar{i}n}}{\bar{i}n}$	$P = n\bar{R}F_{\bar{R}P,\bar{i},n}$ (App. 2, Table 3)
4	$\bar{R}$ for 1 year to P	Present worth of 1 year of unaflow starting X years hence	$P = \bar{R}\, e^{-\bar{i}X}\dfrac{1 - e^{\bar{i}}}{\bar{i}}$	$P = \bar{R}\, F_{SP,\bar{i},X}F_{\bar{R}P,\bar{i},1}$
5	P of flow changing at an exponential rate for n years	Present worth of $R_x = R_0\, e^{\pm \bar{g}X}$ for n years	$P = nR_0\dfrac{1 - e^{-(\bar{i}\mp\bar{g})n}}{\bar{i} \mp \bar{g}}$	$P = nR_0\, F_{\bar{R}P,(\bar{i}\mp\bar{g}),n}$
6	P of flow declining in straight line to zero	Flow goes from R_0 at zero time to 0 in n years; total flow Q is $nR_0/2$	$P = Q\left[\dfrac{2}{\bar{i}n}\left(1 - \dfrac{1 - e^{-\bar{i}n}}{\bar{i}n}\right)\right]$	$P = QF_{SDP,\bar{i},n}$ (App. 2, Table 4)
7	P to $\bar{R}$	Converts a present worth to a unaflow of n years	$\bar{R} = \dfrac{P}{n}\dfrac{1}{(1 - e^{-\bar{i}n})/\bar{i}n}$	$R = \dfrac{P}{n}\dfrac{1}{F_{\bar{R}P,\bar{i},n}}$

Reprinted with permission

Table 1-6 continued

8	$\bar{R}$ to S	Converts a unaflow for n years to a future amount, n years hence	$S = n\bar{R}\, e^{in} \frac{1 - e^{-in}}{in}$	$S = n\bar{R}\, F_{PS,i,n} F_{RP,i,n}$
9	S to $\bar{R}$	Converts a future sum S, n years from now, to a unaflow; sinking-fund payment	$\bar{R} = \frac{Se^{-in}}{n} \frac{1}{(1 - e^{-in})/in}$	$\bar{R} = \frac{S}{n} F_{PS,i,n} \frac{1}{F_{\bar{R}P,\bar{i},n}}$
10	P to K	Converts a present worth representing n years to a capitalized cost	$K = P \frac{e^{in}}{e^{in} - 1}$	
11	K to $\bar{R}$	Converts a capitalized cost to a unaflow	$\bar{R} = iK$	
12	Before to after tax	Converts before-tax amount to after-tax amount, at tax rate t	After tax $= (1 - t\psi)$ before tax	
13	ψ	Present value of \$1 of depreciation, n' years life for tax purpose	$\psi_{SL} = \frac{1 - e^{-in'}}{in'}$ $\psi_{SD} = \frac{2}{in'}\left(1 - \frac{1 - e^{-in'}}{in'}\right)$ $\psi = 1$ for instantaneous tax benefit $\psi = 0$ for no depreciation, i.e., land	$\psi_{SL} = F_{\bar{R}P,\bar{i},n'}$ (App. 2, Table 3) $\psi_{SD} = F_{SDP,\bar{i},n'}$ (App. 2, Table 4)

Jelen, F. C. <u>Cost and Optimization Engineering</u> McGraw-Hill Book Co. (1970)

Reprinted with permission

Table 1-7

Capitalized-cost Relationships with Inflation, $r > d$*

Item no.	*Item*	*Algebraic relationship for capitalized cost*	*Factor relationship for capitalized cost $K_{r>d}$, sum-of-digits depreciation*
1	C_d Depreciable first cost†	$C_d(1 - t\psi_{r_{n'}}) \frac{(1+r)^n}{(1+r)^n - (1+d)^n}$	$C_d(1 - tF_{SDP,r,n'}) \frac{F_{PS,r,n}}{F_{PS,r,n} - F_{PS,d,n}}$
2	R Uniform end-of-year burden†	$R(1 - t) \frac{1+d}{r-d}$	$R(1 - t) \frac{1+d}{r-d}$
3	R_b Uniform beginning-of-year burden†	$R_b(1 - t\psi_{r_1}) \frac{1+r}{r-d}$	$R_b(1 - tF_{SDP,r,1}) \frac{1+r}{r-d}$
4	C_{ex} Irregular cost at end of xth year†	$C_{ex}(1 - t)(1+d)^x \frac{(1+r)^{n-x}}{(1+r)^n - (1+d)^n}$	$C_{ex}(1 - t)F_{PS,d,x} \frac{F_{SP,r,n-x}}{F_{SP,r,n} - F_{SP,d,n}}$
5	C_{bx} Irregular cost at beginning of xth year†	$C_{bx}(1 - t\psi_{r_1})(1+d)^{x-1} \frac{(1+r)^{n-x+1}}{(1+r)^n - (1+d)^n}$	$C_{bx}(1 - tF_{SDP,r,1})F_{PS,d,x-1} \frac{F_{PS,r,n-x+1}}{F_{PS,r,n} - F_{PS,d,n}}$
6	C_{nd} Nondepreciable first cost†	$C_{nd}\,1$	C_{nd}
7	C_{sal} Salvage value†, ‡	$C_{sal}\left[1 + t \frac{(1+d)^n - 1}{(1+r)^n - (1+d)^n}\right]$	$C_{sal}\left(1 + t \frac{F_{PS,d,n} - 1}{F_{PS,r,n} - F_{PS,d,n}}\right)$

* For $d > r$, $K_{d>r} = -K_{r>d}$. For unaburden, $R_D = (r - d)K_{r>d}$ for $r > d$ or $d > r$. For $d = r$ see Sec. 6.12.
† All costs in terms of present dollar.
‡ If excess salvage value due to inflation can be taken as capital gain, use such tax for t in item 7.

can be purchased now for $20,000. How much can be spent now for a new machine which lasts 10 years, with negligible salvage value, and is written off in 10 years using sum-of-year-digits depreciation? Money is worth 16% per year after a 46% tax, and the inflation rate is expected to be 10%.

Unaburden for the 2-year machine from Item 1 Table 1-7 is

$$R_D = (r - d)(C_d)(1 - tF_{SDP,16\%,1})\frac{F_{PS,16\%,2}}{F_{PS,16\%,2} - F_{PS,10\%,2}}$$

$$= (.16 - .10)(10{,}000)\left[1 - .46(.86207)\right]\frac{1.3456}{1.3456 - 1.2100}$$

$$= 3{,}592.9$$

For the same unaburden for the 10-year machine

$$3{,}592.9 = (r - d)(C_d)(1 - tF_{SDP,16\%,10})\frac{F_{PS,16\%,10}}{F_{PS,16\%,10} - F_{PS,10\%,10}}$$

$$3{,}592.9 = (.16 - .10)(C_d)[1 - .46(.58713)]\frac{4.114}{4.114 - 2.5937}$$

$$C_d = \$33{,}804$$

That is, $33,804 could be spent for the 10-year machine. If the inflation rate were zero it would be permissible to spend only $24,891 for the 10-year machine (calculation not shown). Thus inflation adds flavor to the longer lasting article generally.

THE LEARNING CURVE. Productivity increases with time because of technological advancement. Fig. 1-1 shows the direct man hour per lb. of airplane against the cumulative plane number for eight types of fighter planes, with several series in some types, produced by four manufacturers in World War II.

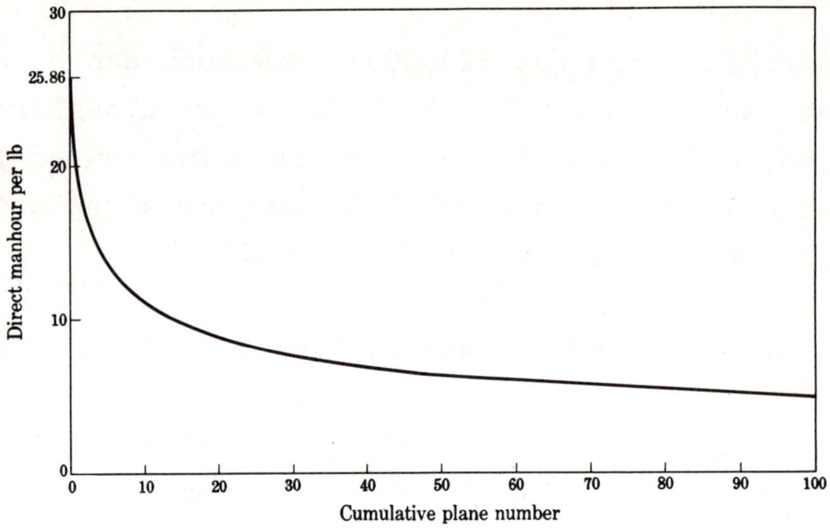

Industry average unit curve for Century service aircraft. (*AMPR data.*)

Fig. 1-1

Jelen, F.C. Cost and Optimization Engineering McGraw-Hill Book Co. (1970)
Reprinted with permission.

If these data are plotted on log-log paper a straight line results as shown in Fig 1-2.

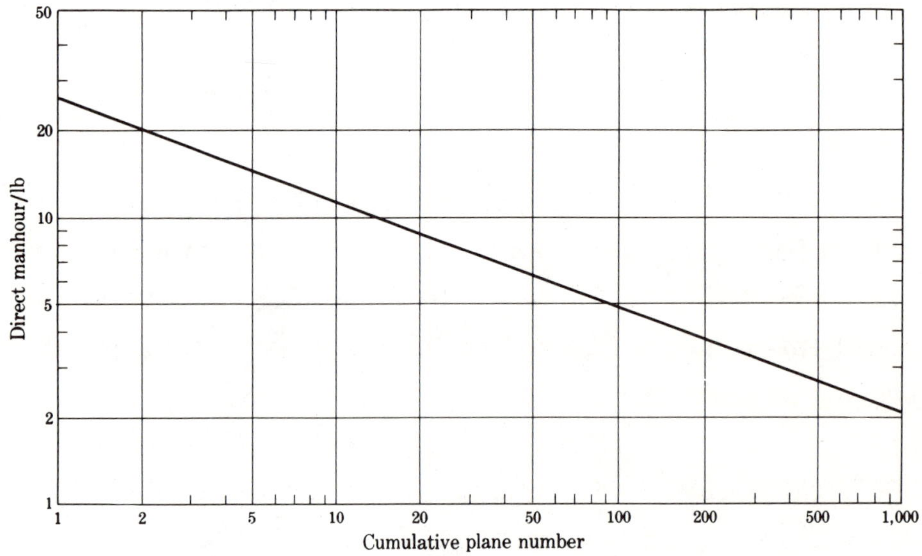

Log-log plot of industry average unit curve for Century series aircraft. (*AMPR data.*)

Fig. 1-2

Jelen, F.C. Cost and Optimization Engineering McGraw-Hill Book Co. (1970)
Reprinted with permission.

This suggests a relationship of the form

$$E_N = KN^s \tag{1}$$

where E_N is the effort per unit of production required to produce the Nth unit. K is a constant, numerically the theoretical effort required to produce the first unit. The slope constant, s, will be negative because the effort per unit decreases with production. Thus

$$\frac{E_2}{E_1} = \frac{K2^s}{K1^s} = 2^s$$

$$\frac{E_4}{E_2} = \frac{K4^s}{K2^s} = 2^s \qquad \text{etc.}$$

Every time production is doubled, the effort per unit required is a constant 2^s of what it was. It is common practice to express the learning curve function in terms of the gain for double production. Thus a 90% learning curve function requires only 90% of the effort per unit every time production is doubled.

If L_p is the percentage learning ratio then the relationship between s, the slope constant, and L_p is given by

$$s = \frac{\log L_p - 2}{\log 2} \tag{2}$$

For the total effort for N units from 1 through N, the cumulative effort E_T is

$$E_T = \sum_1^N E_N$$

An approximation is given by

$$E_T = \frac{K}{s + 1}\left[(N + 0.5)^{s + 1} - 0.5^{s + 1}\right] \qquad (3)$$

The approximation improves as N increases. If N is very large compared to 0.5, the previous equation reduces to

$$E_T = \frac{K}{s + 1} N^{s + 1} \qquad (4)$$

Examples 1-4 and 1-5 demonstrate applications of the Learning Curve

Example 1-4

If 846.2 man-hr is required for the third production unit and 783.0 for the fifth unit, find the percentage learning ratio and the man-hours required for the second, fourth, tenth, and twentieth units.

By Eq. (1)

$E_3 = 846.2 = K(3^s)$

$E_5 = 783.0 = K(5^s)$

By division

$$\frac{846.2}{783.0} = 1.0807 = \left(\frac{3}{5}\right)^s = 0.6^s$$

$$\log 1.0807 = 0.03371 = s \log 0.6 = s(9.77815 - 10) = s(-0.22185)$$

$$s = \frac{0.03371}{-0.22185} = -0.1520$$

By Eq. (2)

$$-(-0.1520) = \frac{2 - \log L_P}{0.30103}$$

$$\log L_P = 1.95424$$

$$L_P = 90.0\% = \text{percentage learning ratio}$$

Using the data for $N = 3$ and $S = -0.1520$

$$846.2 = K(3^{-0.1520})$$

$$\log 846.2 = \log K - 0.1520 \log 3$$

$$2.9278 = \log K - 0.07252$$

$$\log K = 3.0000$$

$$K = 1{,}000$$

Jelen, F. C. Cost and Optimization Engineering McGraw-Hill Book Co. (1970)
Reprinted with permission

so that the learning-curve function is

$$E_N = 1{,}000N^{-0.1520}$$

The effort required for any unit can now be calculated directly. Thus for the twentieth unit

$$E_{20} = 1{,}000(20^{-0.1520})$$

$$\log E_{20} = \log 1{,}000 - 0.1520 \log 20$$

$$= 3.0000 - 0.1520(1.30103) = 2.80224$$

$$E_{20} = 634.2$$

A tabulation for other units is

E_1	1,000.0
E_2	900.0
E_4	810.0
E_5	783.0
E_{10}	704.7
E_{20}	634.2

Note that

$$\frac{E_2}{E_1} = \frac{E_4}{E_2} = \frac{E_{10}}{E_5} = \frac{E_{20}}{E_{10}} = 0.900$$

Example 1-5

Every time production is tripled the unit man-hours required is reduced by 20%. Find the percentage learning ratio.

It will require 80%, or 0.80, for the ratio of effort per unit for tripled production.

By Eq. (1)

$$\frac{E_{3N}}{E_N} = 0.80 = \frac{K(3N^s)}{K(N^s)} = 3^s$$

$$s = \frac{\log 0.80}{\log 3} = \frac{9.90309 - 10}{0.47712} = \frac{-0.09691}{0.47712} = -0.20311$$

By Eq. (2)

$$-(-0.20311) = \frac{2 - \log L_P}{\log 2}$$

$$\log L_P = 1.93886$$

$$L_P = 86.86\% = \text{percentage learning ratio}$$

Jelen, F. C. Cost and Optimization Engineering McGraw-Hill Book Co. (1970)

PROFITABILITY. There are six common criteria for profitability

- Payout time
- Payout time with interest
- Return on original investment (ROI)
- Return on average investment (RAI)
- Discounted cash flow rate of return (DCFRR)
- Net present value (NPV)

Consider a project having an income after taxes but before depreciation (cash flow) as follows

Time, years	Cash flow, kilo$
0	-1,000
0-1	475
1-2	400
2-3	330
3-4	270
4-5	200

and consider the profitability as measured by the six criteria.

Payout time is the time to get the investment back. Working capital does not enter since working capital is always available as such. (The project has no stated working capital anyway). A tabulation gives

Time, end of year	Cumulative cash flow, kilo$
0	-1,000
1	-525
2	-125
3	+205

The initial investment is returned between 2 and 3 years; by interpolation, the payout time is 2.38 years.

Payout time with interest allows for a return on the varying investment. The interest charge may be on the fixed investment remaining only, but another variant applies an interest charge on working capital also. The project leads to the following tabulation with 10% interest

End year	Investment for year	Charge on investment for year at 10%	Cash flow	Cash flow after investment charge	Cumulative net cash
0			-1,000		-1,000
1	1,000	100	475	375	-625
2	625	62.5	400	337.5	-287.5
3	287.5	28.75	330	301.25	+13.75

By interpolation the payout time is 2.95 years allowing for 10% return on the varying investment.

Return on original investment (ROI) regards the investment as fixed and should include working capital. Using straight line depreciation and a 5 year life for the project gives

Year	Cash flow	Depreciation	Profit
1	475	200	275
2	400	200	200
3	330	200	130
4	270	200	70
5	200	200	0
		Average	135

For an original investment of 1,000 kilo$ and no working capital

$$\text{ROI} = \frac{135}{1{,}000 + 0}(100) = 13.5\%$$

The return on average investment RAI is similar to the return on original investment except that the divisor is the average outstanding investment plus working capital. The average investment for the project with straight line depreciation is

Year	Investment, kilo$	
1		1,000
2	1,000 - 200 =	800
3	800 - 200 =	600
4	600 - 200 =	400
5	400 - 200 =	200
	Average =	600

$$RAI = \frac{135}{600} 100 = 22.5\%$$

The discounted cash flow rate of return considers the timing of all cash flows. The method finds the rate of return that makes the present worth of all the receipts equal to the present worth of all the expenses. For the project under study

$$P = 0 = \frac{-1,000}{(1 + r)^o} + \frac{475}{1 + r} + \frac{400}{(1 + r)^2} + \frac{330}{(1 + r)^3} + \frac{270}{(1 + r)^4} + \frac{200}{(1 + r)^5}$$

By repeated trials

$$r = DCFRR = 23.9\%$$

The net present value (NPV) is calculated without a trial and error calculation and is the present value calculated using a given rate of return, say the minimum acceptable rate of return. For the project at 10%

$$NPV = \frac{-1,000}{(1.10)^o} + \frac{475}{(1.10)} + \frac{400}{(1.10)^2} + \frac{330}{(1.10)^3} + \frac{270}{(1.10)^4} + \frac{200}{(1.10)^5}$$

$$NPV = 318.92$$

A simple format for calculating cash flow and from it the DCFRR is given in Table 1-8. Depreciation is not an out of pocket expense but it must be considered in calculating the taxable income (tax base) on which taxes are paid. Table 1-8 is for a venture that requires $1 million now and will have receipts of 0.6, 0.7, and 0.34 Mega$ at the end of the first, second and third years respectively. Consider these receipts as the difference between income less expenses but before taxes and depreciation. The combined federal and state income tax is 55% and sum-of-the-years-digits depreciation is used. As shown for zero present worth of the cash flows, the DCFRR is 15.8% by interpolation.

Table 1-8 **Cash Flow for Example**

	Time 0	*End year* 1	*End year* 2	*End year* 3	*Cumulative*
1. Total receipts		600,000	700,000	340,000	
2. Depreciation					
$\frac{3}{1+2+3}$ 1,000,000		500,000			
$\frac{2}{1+2+3}$ 1,000,000			333,333		
$\frac{1}{1+2+3}$ 1,000,000				166,667	
3. Taxable income		100,000	366,667	173,333	
4. Tax at 55%		55,000	201,667	95,333	
5. Profit after tax		45,000	165,000	78,000	288,000
6. Cash flow, item 5 + item 2	−1,000,000	545,000	498,333	244,667	288,000
DCF rate of return, discrete compounding					
Present worth at 14%	−1,000,000	478,069	383,450	165,143	+26,662
Present worth at 16%	−1,000,000	469,828	370,341	156,748	−3,083
For 0 present worth $r = 15.8\%$ by interpolation					

Reprinted with permission from Jelen, F.C. Cost and Optimization Engineering McGraw-Hill Book Co. (1970)

COST ESTIMATION is included in Part 3 of the certification examination. To avoid needless duplication in the book at hand material on cost estimation has been collected in Chapter 3. Since those taking the examination will have varied experience, questions in the examination will be somewhat general; but the applicant should be will versed in the fundamentals of cost estimation and their applications.

COST CONTROL. Chapter 17 in "Cost and Optimization Engineering" has a good summary of cost control and several check-lists. Chapter 5 of the book at hand has additional information. The student should be fluent with these two sources when taking the examination.

The critical path method CPM is an essential part of construction and is now required in many contracts. The student should be familiar with the terms used in CPM, the essentials of network diagrams, and their use and advantages.

WORKING CAPITAL comprises the funds, in addition to fixed capital and start up expenses, needed to get a project started and meet subsequent obligations. For a manufacturing plant the working capital includes

- Raw materials inventory
- Work in progress inventory (semifinished goods)
- Finished product inventory
- Supplies for product manufacture
- Accounts receivable
- Cash on hand for salaries, wages, etc.
- Accounts payable

Note that accounts receivable adds to the working capital required while accounts payable reduces the working capital required. In general working capital is based on the above items for a one month period. In manufacturing industries, working capital is about 10 to 20% of fixed capital investment. In some industries like those in sales with a high inventory of finished goods, the figure can become 50% and higher. For a rough extimate, working capital may be taken as 15% of fixed capital plant investment.

CONCEPTUAL COSTING for a project involves estimating the cost of a project with very little data. It is not required necessarily that a flow sheet be available. It is best adapted to the use of basic cost data from a project with similar processing. For example, basic project cost data for chlorination of benzene might be used for conceptual costing of a project involving similar chlorination of methyl benzene (toluene). Other examples might include hydrogenation, sulfonation, oxidation, alkylation and polymerization. In practice several parameters might have to be included to achieve any degree of accuracy. Thus fixed capital costs for polymerization can vary widely. Vinyl chloride polymerizes easily and ethylene polymerizes with difficulty. Here "polymerization" of itself cannot be classified as a definite capital cost. Conceptual costing may be used for

- Ascertaining project profitability in its early development.
- Quickly checking bids from contractors and subcontractors.
- Estimating costs and profits for potential process licensing.
- Quick accessment of inquiries for increased product sales.

PROJECT PLANNING AND SCHEDULING. Material on these topics is collected in Chapter 6. Participants in the examination should be well versed with these topics.

COST ACCOUNTING. Material on cost accounting is assembled in Chapter 2.

SOURCES OF DATA. The cost engineer should be very familiar with sources of data. Table 1-9 gives some general sources.

FORECASTING. Mathematical methods for extrapolating trend curves are shown in Fig 1-3.

COST INDICES are discussed in Chapter 7.

VALUE. Value and cost are not the same concept. The nature of value is discussed in Table 1-10. There is an organization known as SAVE--Society of American Value Engineers. As the concepts of cost engineering and value engineering each broaden, it is likely that the two will have much common interest in the future.

Table 1-9

SOURCES OF DATA

The most common sources of economic data include government agencies, trade associations, and business publications. When no published data are available, special tabulations or surveys can be made and in some cases may be warranted.

Much useful economic data can be found in readily available publications. The following list of information sources is not meant to be all-inclusive, but a familarity with the references included in this list will prove extremely useful to anyone concerned with economic statistics.

Directories

Statistics Sources. Gale research, Detroit, Mich.
Executive's Guide to Information Sources. Gale Research, Detroit, Mich.
Business Trends and Forecasting Information Sources. Gale Research, Detroit, Mich.
Encyclopedia of Associations. Gale Research, Detroit, Mich.
Thomas Register of American Manufacturers. Thomas, New York.
Poor's Directory of Executives. Standard & Poor, New York.
Million Dollar Directory. Dun & Bradstreet, New York.

General Economic Data

Statistical Abstract of the United States. U.S. Government Printing Office, Washington, D. C. (annual).
Economic Almanac. National Industrial Conference Board, New York (annual).
Survey of Current Business. U.S. Office of Business Economics, Washington, D. C. (monthly).
National Economic Projections. National Planning Association, Washington, D. C.
Economic Indicators. U.S. President's Council of Economic Advisors, Washington, D. C. (monthly).
Handbook of Basic Economic Statistics. Economic Statistics Bureau, Washington, D. C. (monthly).

Industry and Product Data

U.S. Industrial Outlook. U.S. Bureau of Defense Services Administration, Washington, D. C. (annual).
Predicasts. Predicasts, Cleveland (quarterly).
Census of Business. U.S. Bureau of the Census, Washington, D. C.
Census of Manufactures. U.S. Bureau of the Census, Washington, D. C.
Census of the Mineral Industries. U.S. Bureau of the Census, Washington, D. C.
Census of Population. U.S. Bureau of the Census, Washington, D. C.
Minerals Yearbook. U.S. Bureau of Mines, Washington, D. C. (annual).
Construction Review. U.S. Bureau of Defense Services Administration. Washington, D. C. (monthly).

Area Information

Commercial Atlas and Marketing Guide. Rand-McNally, Chicago.
Editor and Publisher Market Guide. Editor and Publisher, New York (annual).
Survey of Buying Power. Sales Management Magazine, New York (annual).
Municipal Yearbook. International City Manager Association, Chicago (annual).

Financial Data

Quarterly Financial Report for Manufacturing Corporations. U.S. Federal Trade Commission and U.S. Securities and Exchange Commission, Washington, D. C. (quarterly).
Leading U.S. Corporations. News Front Magazine, New York.
Moody's Handbook of Common Stocks. Moody's Investors Service, New York (quarterly).
Stock Guide. Standard & Poor, New York (monthly).
Value Line Investment Survey. Arnold Bernhardt, New York (weekly).

General Business and Financial Periodicals

Wall Street Journal. Dow Jones, New York (daily).
Business Week. McGraw-Hill, New York (weekly).
Forbes. Forbes, New York (bimonthly).
Fortune. Time, New York (monthly).
Dun's Review. Dun & Bradstreet, New York (monthly).

In predicting the future, then, it is necessary to "know the past and understand the present." A forecast of economic activity in any business or industry involves an analysis of past trends, extrapolation of these trends into the future, and adjustment, refinement, or revision of the projected trends to allow for any factors expected to cause deviations from the established trends.

The specific techniques to be employed in economic forecasting depend largely upon how far into the future the forecast extends. For extremely short-range forecasts—perhaps covering only a week or a month—it may be safe to assume that the immediate future will be very much like the present, or at least that recent trends will continue for a while. Using this approach proves adequate most of the time, significant errors occurring only when there is a major reversal of trends.

A second method of forecasting, frequently used for intermediate-term forecasts, employs "leading" indicators to predict coming changes. Next year's construction activity, for example, will reflect this year's contract awards. Some of the most important indicators used by economists include interest rates, contract awards, stock prices, manufacturers' new orders, freight car loadings, indexes of industrial production, corporate profits, and business failures.

Fig. 1-3 Extrapolation of curves for forecasting

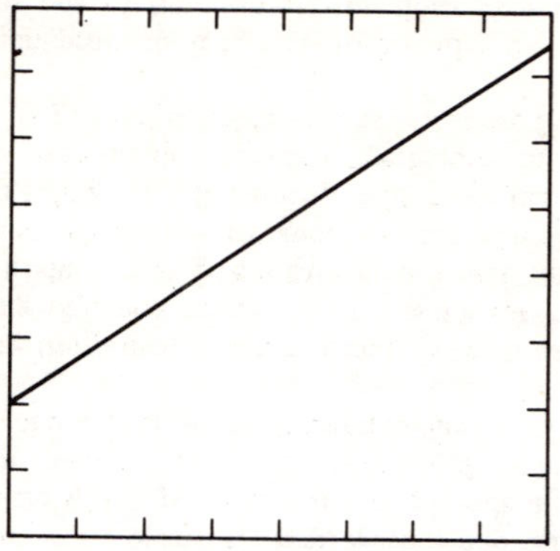

$y = a + bx.$

Arithmetic trend.

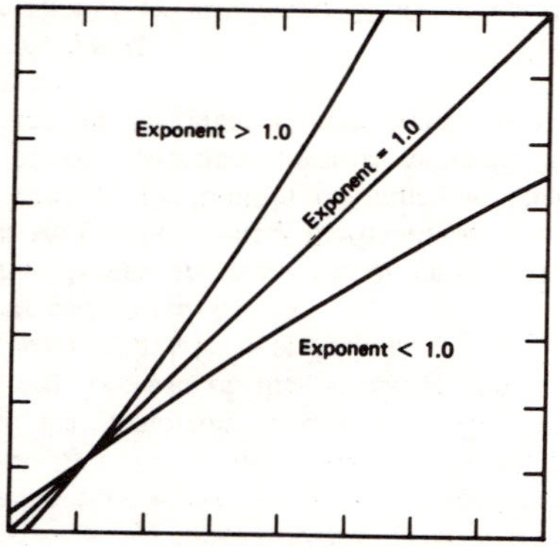

$y = ax^b.$

Logarithmic trend.

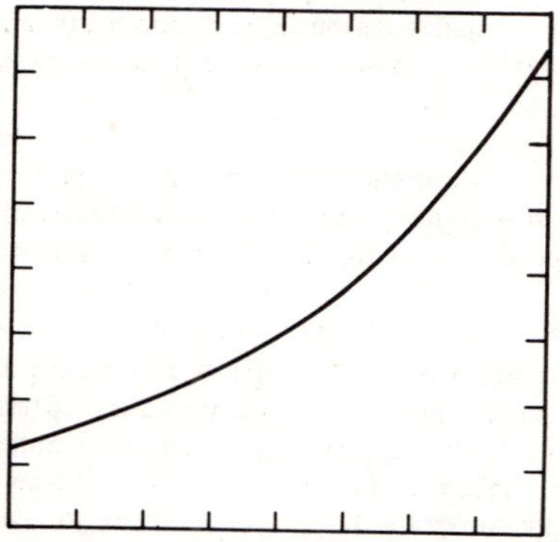

$y = ab^t.$

Semilogarithmic trend.

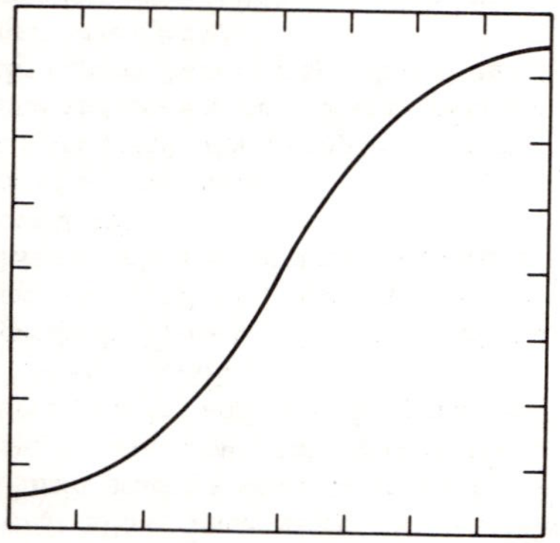

S trend.

THE TREND. The S-shaped curve plots cumulatively a "normal" or binomial frequency distribution of data. This type of curve is especially valuable in forecasting future sales of a specific product, since the life-cycles of most products generally follow this shape. They typically start slow, pick up momentum as they gain market acceptance, and then taper off gradually as the market becomes saturated and newer products begin replacing them. A symmetrical S curve plots as a straight line on probability graph paper.

Reprinted with permission of John Wiley and Sons from W. R. Park <u>Cost Engineering Analysis</u> (1973)

Table 1-10

What is value?

The concept of value is something which we all apply in our everyday lives when we part with money to acquire an item of furniture or a washing machine or the weekly supply of groceries. Value is not necessarily represented by the lowest-priced product in its class but rather by the one which, for the purposes for which we want it, gives a better combination of ideas, functions, reliability, etc., for its price than the competition.

Value is not an inherent property of a product. It can only be assessed by comparison with other similar products on offer at the same time and under the same conditions. It can be considered in four types:

1 *Exchange value:* that which enables an article to be offered in exchange for money or for another article.
2 *Cost value:* the sum of the materials, labour and other costs required to produce an article or to perform a service.
3 *Use value:* the sum of the properties and qualities possessed by an article which perform a use or provide a service.
4 *Esteem value:* the sum of the features of the article which, beyond its actual use, prompt the decision to buy.

In industry the principal concern must be with use value, the highest value being obtained where the cost is lowest within the limitation that the product must give a satisfactory performance. However, the VA exercise which tries to ignore esteem value or to eliminate it is very often riding for a fall since it will, more than likely, recommend a product which, although near perfection in respect of its performance, is just not saleable – it has lost those features which attract the buyer.

In certain products it is vital to consider the provision of esteem features as one of the essential functions. A water tank which will be hidden in the roof space of a house needs no features other than those concerned with its use but a cooker to be installed in the kitchen needs many features which have nothing to do with its use or the ladies will never accept it! In summary, good value is the lowest cost reliably to accomplish the essential functions.

Unnecessary cost

Value analysis is an unrelenting search for waste or, put in other words, it is an organised campaign for the elimination of unnecessary cost. Unnecessary cost is defined as any expenditure which can be removed without impairing the quality, reliability, saleability or (if applicable) the maintainability of the subject.

But what allows unnecessary cost to appear? The reasons fall into six main headings which must not be thought to be critical of the design function alone, but are often just as applicable to buying, production, planning and, indeed, to management. They are:

1 *Lack of information.*
Is everbody up-to-date in respect of materials, methods, techniques, etc? And, of prime importance, is accurate, detailed and understandable cost information available outside the enclave of accountancy?

2 *Lack of original ideas.*
What are people doing with their brains? Using them creatively to find and develop new ideas or reducing them to mere memory banks capable only of responding to the dictates of tradition?

3 *Honest wrong beliefs*
Long-established and honestly-held beliefs can be wrong, even on technical matters, and can lead to unnecessary costs.

4 *Temporary or changed circumstances and time pressures.*
A temporary expedient is introduced 'to-get-it-into-production'. 'It's costly but don't worry, old man, it will be put right tomorrow' – and that tomorrow never comes! Or – 'We shall have to do it this way because of the particular conditions of this order – it will be put right later' – or will it?

5 *Habits and attitudes.*
Human beings are creatures of habit and adopt attitudes. The older we get the stronger the habits become and, by consequence, the stronger the built-in resistance to change. To what extent are personal attitudes towards materials, methods, money and, perhaps, of the greatest importance, men, allowed to influence decisions?

6 *Over-designing.*
This should be considered in its broadest sense and not confined to the designer. Does the specification give the customer exactly what he wants? He may want something much more simple. If so, he is getting poor value.

LINEAR PROGRAMMING--SIMPLEX METHOD is important and is discussed more fully in Chapter 8, but a simple problem is illustrated here in a concise manner. Two types of products A and B can be made and each must go through Machine bank 1 and Machine bank 2. Machine bank 1 is limited to 6,000 hr/week and Machine bank 2 is limited to 10,000 hr/week. One item of A requires 2 hr on machine type 1 and 1 hr on Machine type 2. One item of B requires 1 hr on Machine type 1 and 4 hr on Machine type 2. Find the policy that will maximize the profit per week. (A = \$3.5, B = \$5)

The Simplex Method is an algorithm that leads to a best solution by starting with a feasible solution and improving it until no further improvement is possible. Let x_1 be the number of A items made per week and x_2 the number of B items made per week. Then the restrictions are

$$2x_1 + x_2 \leq 6{,}000$$
$$x_1 + 4x_2 \leq 10{,}000$$

Since mathematics cannot handle inequalities very well, we introduce "slack" variables x_3 and x_4 which are positive in amount, that is here use up machine time, but here have no monetary value in the profit. The above equations then become equalities and are

$$2x_1 + x_2 + x_3 = 6{,}000 \qquad (5)$$
$$x_1 + 4x_2 + x_4 = 10{,}000 \qquad (6)$$

and the objective function is to maximize

$$Z = 3.5x_1 + 5x_2 \qquad (7)$$

With 3 equations and 4 unknowns, there are an infinite number of solutions, but only one solution will give a maximum for Z. The Simplex Method finds that solution.

Start with a feasible mix. Here it would be feasible, but

not optimum, to make neither A nor B, that is, use idle time x_3 and x_4 only. The first tableaux of the Simplex Method will be

	W_j →	3.5	5	0	0		
	Item →	X_1	X_2	X_3	X_4	B_i	
Item	W_j						
X_3	0	2	1	1	0	6000	$\frac{6000}{1} = 6000$
X_4	0	1	(4)	0	1	10000	$\frac{10000}{4} = 2500$ ↗ smallest positive
	Z_j	0	0	0	0	0	
Z_j -	W_j	-3.5	-5 ↑ smallest negative	0	0		

Note that the row 2 1 1 0 6000 is merely Equation (5) and the row 1 4 0 1 10000 is Equation (6). The Z_j in each column is the sum of $(W_j)(X_j)$. Thus the succesive values for Z are

$$(0)(2) + (0)(1) = 0$$
$$(0)(1) + (0)(4) = 0$$
$$(0)(1) + (0)(0) = 0$$
$$(0)(0) + (0)(1) = 0$$
$$(0)(6000) + (0)(10000) = 0$$

The arrow points to -5, the smallest *negative* in the last row. This is the pivot column. The very last column is obtained by dividing B in each row by the number in the same row of the pivot column. The numbers are 6000/1 = 6000 and 10000/4 = 2500. The smallest positive is 2500 which indicates the pivot row. The number common to the pivot row and pivot column is circled and is the pivot element.

The pivot element is now made equal to unity, here dividing by 4, to give the corresponding row in a second tableaux. Thus for the second tableaux row 2 becomes

$$\frac{1}{4} \quad 1 \quad 0 \quad \frac{1}{4} \quad 2500 \qquad (8)$$

All other elements in the pivot column must be made equal to zero. Here Multiply row 1 of the first tableaux by -1 and add it to Equation (8).

-2	-1	-1	0	-6000
$\frac{1}{4}$	1	0	$\frac{1}{4}$	2500
$\frac{-7}{4}$	0	-1	$\frac{1}{4}$	-3500

The last line becomes the first row of the second tableaux. Also x_4 in the pivot row is replaced by x_2 in the pivot column. The second tableaux becomes

		3.5	5	0	0	B_i
		X_1	X_2	X_3	X_4	
X_3	0	$-\frac{7}{4}$	0	-1	$\frac{1}{4}$	-3500
X_2	5	$\frac{1}{4}$	1	0	$\frac{1}{4}$	2500
	Z_j	$\frac{5}{4}$	5	0	$\frac{5}{4}$	12500
Z_j -	W	$-\frac{9}{4}$	0	0	$\frac{5}{4}$	

The procedure is repeated leading to a third tableaux (details not shown).

		3.5	5	0	0	
		X_1	X_2	X_3	X_4	B_i
X_1	3.5	1	0	$\frac{4}{7}$	$-\frac{1}{7}$	2000
X_2	5	0	1	$-\frac{1}{7}$	$\frac{2}{7}$	2000
	Z_j	3.5	5	$\frac{9}{7}$	$\frac{13}{14}$	17000
Z_j -	W	0	0	$\frac{9}{7}$	$\frac{13}{14}$	

There is no negative number in the last row, hence, no improvement is possible. The best policy is to make 2000 of X_1 (A) and 2000 of X_2 (B) with a profit of 17000 $/week.

DYNAMIC PROGRAMMING is useful for finding the optimum policy for a problem involving steps or stages. It can handle non-linear relationships. The essence of the technique is to reduce the number of policies that must be assessed to find the optimum policy.

Consider the following problem. A salesman must travel from city A to city B and he must visit one city in each of the three districts shown in Fig. 1-4. Travel costs between the cities are shown in the Figure.

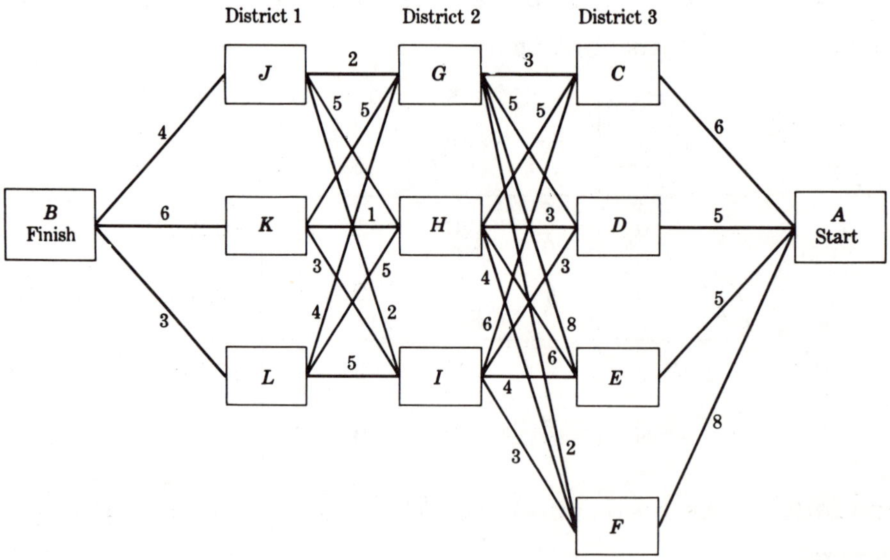

Fig. 1-4 Travel Costs

Reprinted with permission from Jelen, F.C. Cost and Optimization Engineering McGraw-Hill Book Co. (1970)

Find the policy for minimum travel cost. Here the total number of possibilities is only 4 x 3 x 3 = 36 but if there were 10 districts with 10 cities the number of possibilities would be

10^{10}, a formidable number for even a computer. Dynamic programming is a systematic method which greatly reduces the number of trials in a large problem.

Consider destination city B. There are 3 routes from District 1 to B, namely JB, KB, and LB. Having arrived in District 1 there is no decision to be made and the costs are

JB	4	KB	6	LB	3

Go back one stage to District 2. All the combinations for going from District 2 to B are

GJB	6	HJB	9	IJB	6
GLB	11	HKB	7	IKB	9
GLB	7	HLB	8	ILB	8

Nothing has been gained in the reduction of trials to this point, but from here on reduction can be realized. The 9 routes just listed can be reduced to the best from G, from H, and from I. They are

GJB	6	HKB	7	IJB	6

Only these 3 (instead of the previous 9) need to be considered in the tabulation for District 3 to City B, which becomes

C-GJB	9	D-GJB	11	E-GJB	14	F-GJB	8
C-HKB	12	D-HKB	10	E-HKB	13	F-HKB	11
C-IJB	12	D-IJB	9	E-IJB	10	F-IJB	9

These 12 routes can be reduced to only 4, namely

CGJB	9	DIJB	9	EIJB	10	FIJB	8

Finally the choice is that for

A - CGJB	15
A - DIJB	14
A - EIJB	15
A - FGJB	16

Minimum cost is $14 for route ADIJB.

Dynamic Programming is discussed more fully in Chapter 8.

REFERENCES

1. Abdel-Aal, H.K. and Schmelzlee, R., Petroleum Economics and Engineering, Marcel Dekker, Inc., (1976).
2. American Association of Cost Engineers, Cost Engineers' Notebook , A continuing publication.
3. American National Standards Institute, ANSI.
   "Cost Engineering Terminology", Z94.2
   "Distribution and Marketing Terminology", Z94.4
   "Engineering Economy Terminology", Z94.5
   "Facility Planning Terminology", Z94.6
   "Materials Processing Terminology", Z94.7
   "Organization Planning and Theory Terminology", Z94.9
   "Production Planning and Control Terminology", Z94.10
   "Work Measurement and Methods Terminology", Z94.12
   1972 Index to, Z94.1 - Z94.12
4. Antill, J.M. and Woodhead, R.W., Critical Path Methods in Construction Practice , 2nd Edition. Wiley Interscience - A Division of John Wiley and Sons, (1970).
5. Baker, K.R., Introduction to Sequencing and Scheduling , John Wiley and Sons, (1974).
6. Bonney, J.B. and Frein, J.P., Handbook of Construction Management and Orientation , Van Nostrand Reinhold Co., (1973).
7. Brown, R.G., Smoothing, Forecasting and Prediction , Prentice-Hall, (1963).
8. Collier, K., Fundamentals of Construction Estimating and Cost Accounting , Prentice-Hall, (1974).
9. Dell' Isola, A.J., Value Engineering in the Construction Industry Construction Publishing Co., (1973).
10. Diez, R.L. and Maloney, L.D., Builders Estimating Factbook , Cahners Books International, (1974).
11. Dodge Building Cost Services, Dodge Manual - For Building Construction Pricing and Scheduling , McGraw-Hill Information Systems Co., current edition.
12. Douglas, C.J. and Munger, E.L., Construction Management , Prentice-Hall, (1969).
13. English, J.M., Editor, Cost Effectiveness - The Economic Evaluation of Engineered Systems , John Wiley and Sons, (1968).
14. English, J.M., Editor, Economics of Engineering and Social Systems , Wiley Interscience, A division of John Wiley and Sons, (1972).
15. Godel, J.B., Guide to Information Science in the Construction Industry , Construction Publishing Co., (1975).
16. Guthrie, K.M., Process Plant Estimating, Evaluation, and Control , Craftsman Book Company of America, (1974).
17. Heller, E.D., Value Management - Value Engineering and Cost Reduction , Addison - Wesley Publishing Co., (1971).
18. Jelen, F.C., Editor, Cost and Optimization Engineering , McGraw-Hill Co., (1970).

19. Land, C.B., Industrial Machinery and Equipment Pricing Guide , Construction Publishing Co., Current Edition.
20. Lock, D., Engineer's Handbook of Management Techniques, Gower Press, Ltd., (1973).
21. Ludwig, E.E., Applied Project Management for the Process Industries , Gulf Publishing Co., (1974).
22. McKee-Berger-Mansueto, (compiled by) Building Cost File . Southern, Central, Western and Eastern Editions. Construction Publishing Co., current edition.
23. McKee-Berger-Mansueto, (compiled by), Design Cost File, 1975 Composite Prices , Construction Publishing Co.,
24. Morris, W.T., Engineering Economic Analysis , Reston Publishing Co., (1976).
25. Nickerson, C.B., Accounting Handbook for Nonaccountants , Cahners Publishing Co.
26. Page, J.S., General Construction Man-hr Manual, Piping Man-hr Manual, Manual of Equipment and Installation Cost, Man-hr Manual, Heating, Air Conditioning, Ventilation, Plumbing, Equipment Installation Man-hr Manual , Gulf Publishing Co.
27. Park, W.R., Cost Engineering Analysis , Wiley Interscience, John Wiley and Sons, (1973).
28. Pilcher, R., Appraisal and Control of Project Costs , McGraw-Hill Book Co., (1973).
29. Poage, S.T., Quantitative Management Methods for Practicing Engineers , Barnes and Hoble, (1970).
30. Radcliffe, B.M., Kawal, D.E., and Stephenson, R.J., Critical Path Method , Cahners Publishing Co., (1967).
31. Revsine, L., Replacement Cost Accounting , Prentice-Hall, (1973).
32. Rose, L.M., Engineering Investment Decisions (planning under uncertainty). Elevier Scientific Publishing Co., (1976).
33. Specthrie, S.W., Industrial Accounting , 2nd Edition. Prentice-Hall, (1959).
34. Volpe, S.P., Construction Management Practice , Wiley Interscience, John Wiley and Sons, (1972).
35. Walker, F.R., Building Estimator's Reference Book , Frank R. Walker Co., current edition.
36. Wells, G.L., Process Engineering With Economic Objective , John Wiley and Sons, (1973).
37. Wheelwright, S.C. and Makridakis, S., Forecasting Methods for Management , Wiley Interscience, John Wiley and Sons, (1973).
38. Wilson, R.M.S., Cost Control Handbook , John Wiley and Sons, (1975).
39. Woods, Donald R., Financial Decision Making in the Process Industry , Prentice-Hall, Inc., (1975).

2. COST ACCOUNTING

Wilfred Stelly and A. J. Wallace
CE-Lummus, Houston, Texas

INTRODUCTION. The cost engineer must be familiar with cost accounting since it is the cost accountant who keeps the cost records. Moreover, the cost engineer should be part of a team in the allocation of overhead and other indirect costs. The subject of cost accounting is covered in considerable detail by Dr. E. H. Weinwurm in Chapter 19 of Dr. Jelen's book on "Cost and Optimization Engineering".

OUTLINE. The chapter is composed of two parts. The first part discusses the basic concepts of cost accounting and how they are used on a project. The second part dicusses various classifications of account. A summary follows in Table 2-1

Table 2-1. Summary of Chapter 2

Part 1. Cost Accounting
- Estimated project sales (selling price)
- Estimated cost of the work (total base cost)
- Gross profit (total project overhead and profit)
- Overhead costs
- Profit or loss before taxes
- Role of cost accounting and definitions
- Chart of accounts

Part 2. Classifications of accounts
- Direct costs
- Indirect costs
- Overhead costs
- Standard costs, budgets, and variances
- Joint and transfer costs, and pricing
- Costing products and services

The presentation in the first part of this chapter favors work associated with engineering and construction for the oil and chemical industries to match the author's experience, but the principles are general.

PROJECT COST ACCOUNTING. In order to understand the function of cost accounting the terms "total cost" and "profit" must be understood. For a business to be successful it must earn a profit. Refer to Table 2-1. Several terms for sales are employed by accountants like: selling price, contract value, or billable value. They represent the total expected income from the customer.

Considered next are the estimated costs of producing the product be it manufactured or a service. Many firms use the term "total base cost" and have commonly a standard list which defines all the elements which account for the cost of the work.

Subtracting the total base cost from the estimated project sales or income gives what is commonly termed "gross profit" or "total overhead and profit." Another term used is "gross margin."

Considered next are the costs involved in selling the product or services. The most common terms used are "overhead cost" or "selling and general administration expenses." These are elements of overhead costs which are discussed in more detail below. After deduction of the overhead costs there remains the profit or loss before taxes. Taxes are discussed in Chapter 1 .

ROLE OF COST ACCOUNTING AND TERMS. Cost accounting, the principal objective of this chapter expands the techniques of financial accounting, and is part of managerial accounting as opposed to historical or audit functions. A general definition of accounting is (Cost and Optimization Engineering, Chapter 19):

"Accounting is the art of recording, classifying, and summarizing in a significant manner and in terms of money, transactions and events which are, in part at least, of a financial character, and interpreting the results thereof."

All accounting systems include three basic steps, recording, classifying, and summarizing economic data in terms of money. Classification of accounts is based on a listing called the "chart of accounts" which plays a major role in all accounting systems. In todays business it is employed extensively in computer applications in the form of management information systems. A good chart of accounts structure accomplishes the following.

- A standard method upon which a business prepares cost estimates in a consistent manner
- A means for recording and classifying cost to permit direct comparison with estimates and budgets
- Facilitation for creation of cost centers such as department or sections into smaller segments for ease of control and obtaining unit return cost data
- The opportunity for the cost engineer to follow the chart of accounts in trending the project costs and in preparing cost forecasts and management cost reports.

The cost engineer should be completely familiar with his company's chart of accounts.

Before considering various classifications of cost, some cost definitions and a few additional accounting terms are examined briefly. Thus for cost (Cost and Optimization Engineering, Chapter 19).

"Cost is the amount, measured in money, or cash expended or other property transferred, capital stock issued, services performed or a libility incurred, in consideration of goods or services received."

ACCOUNTING TERMS. Familiarity with terms used by cost accountants is required. Examination question could include some of the terms as a test of general understanding. The most commonly used terms are given in Table 2-2.

Table 2-2. Accounting terms

·Debit	·By product	·Indirect costs
·Credit	·Process costs	·Cost of sales
·Assets	·Prepaid costs	·Variable costs
·Liabilities	·Standard costs	·Burden rate
·Manufacturing expenses	·Accounts payable	·FIFO
	·Receivables	·LIFO
·Cost of goods sold	·Fixed costs	·Gross margin
·Joint costs	·Direct costs	·Unit cost

Accounting uses a double-entry system for every transaction always with valuation in terms of money. In engineering for no accumulation

$$\text{input} = \text{output}$$

so in accounting

credit = debit

The terms debit and credit are conventional and have no particular significance as such. The distinction between the terms can always be resolved by reduction to a transfer of money between accounts with the accountant acting as an intermediate who transfers the value from one account to another. Suppose a company pays a supplier $500 cash that is owed to the supplier. The account for the company records a balanced transaction as follows

Supplier	Debit $500
Cash	Credit $500

Here the supplier's account is a debit since $500 was put into the supplier's account. Similarly the cash account is a credit since $500 was obtained from the cash account.

Two basic classes of accounts must be distinguished.

- Real accounts. Real accounts are allowed to accumulate indefinitely and are not closed out at the end of the year.
- Revenue and expense, or nominal accounts. They are cleared into capital at the end of the year.

Real accounts are of two types. If they are owned by the business they are called assets. Conversely those owed by the business are called liabilities. In accordance with the double entry equation

debits = credits
assets = liabilities & ownership

where ownership represents the owner's equity in the business.

The meaning of terms used in cost accounting vary from company to company. Hence it is necessary to know exact meaning of terms as used by the cost engineer's company as well as the general meaning. For example the term "manufacturing expense" to some accountants should not include factory overhead. Factory overhead is part of product cost and will funnel into the expense stream only when the product costs are released as cost of goods sold.

"Cost of goods sold" is a widely used term that is somewhat misleading in connection with the meaning of cost. Cost of goods sold is an expense because it is an expired cost, and is as much as expense as are salesmen's commissions. Cost of goods sold is often called "cost of sales."

The distinction between direct costs and indirect costs can cause confusion. The topic is discussed more fully below.

Chapter 2. Part 2. Classification of Cost

INTRODUCTION. Costs can be classified as unexpired and expired as summarized in Table 2-3.

Unexpired costs (assets) are those which are applicable to the production of future revenues. Expired costs are those which are not applicable to the production of future revenues and for that reason, are treated as deductions from current revenues or are charged against retained earnings.

Some examples are

1. Equipment and machinery. This could be automobiles,

Table 2-3. Unexpired and expired costs

Unexpired costs-future revenues
- ·1. Equipment and machinery
- ·2. Inventory and raw materials
- ·3. Prepaid costs
  - Insurance
  - Taxes
  - Rent
- ·4. Salaries

Expired costs-current revenues
- ·1. Depreciation expenses
- ·2. Raw materials (direct materials)
- ·3. Write-off of prepaid costs
- ·4. Salaries

trucks, desks, typewriters, or machines in a machine shop. All are assigned an estimated useful life, and they are written off or depreciated over a given period of time. The useful life of any piece of equipment is determined by government regulation or by guidelines set by historical data. The guidelines are flexible and may be changed from time to time by modern technology.

When the write-off is made, the cost becomes an expired cost. The write-off or depreciation, depending on what it is for, can be classified as direct, indirect, or overhead cost.

2. Raw materials inventory. An accumulation of unused items or components (raw materials) related to producing a product is called an inventory. The value or cost of an inventory may have been acquired at different times and at different prices. In such cases a rule or guideline is required to determine the value at which these units shall be transferred out of inventory and into expired costs. The rule applies to pricing and has nothing to do with which unit is removed physically from the inventory.

A common method of inventory costing is FIFO or first in-first out. The cost of the oldest cost in inventory is used. Under conditions of increasing prices or inflation, lower acquisition prices are matched with higher selling prices, resulting in higher accounting profit figures and higher income taxes. To avoid these effects a method of evaluation can be used known as LIFO or last in-first out which matches current costs with current revenues. However, the lower cost items remain in inventory. The use of an average valuation represents an effort to find a compromise between the two methods.

Many companies have changed to the LIFO method because of the current inflationary spiral.

3. Prepaid costs. These are costs for which cash has been expended for a service or benefit that extends over more than one production cycle, or one year. For example, a three-year $300 insurance premium may be charged originally to an asset account-Prepaid Costs. Subsequent accounting for the cost will hinge on (1) the amount applicable to the current period, say $100 for the first year, and (2) the purpose of the insurance coverage.

Insurance on factory machinery is inventorial and is, therefore, transferred from Prepaid Insurance to an inventory account. Insurance on a sales office is not inventorial and is, therefore, transferred from Prepaid Insurance to an outright expense account.

The same determination has to be made for local taxes and rent.

4. Salaries. They can be an unexpired cost. Work on a particular product could be done, perhaps, months before the final assembly of the items is completed. Also, a portion of the product could be fabricated outside and returned for final assembly. The salary is then written off to the job or products. The salary is written off to the job or products in the end. Unexpired costs are found in balance sheets, and expired costs are found in the profit and loss (income) section of the financial statements.

Cost data, however, found in conventional financial statements are for the specific purpose of reporting to interested outsiders and are not necessarily useful for other applications. The requirement of selecting the appropriate cost type for individual objectives is important. No general rules are possible because of the wide variety of circumstances.

CLASSIFICATION OF COSTS. Table 2-4 gives a summary.

Table 2-4. Classification of costs

- Direct costs
- Indirect costs
- Overhead costs
- Standard costs, budgets, and variances
- Joint and transfer costs, and pricing
- Costing products and services

DIRECT COSTS are also called prime costs and are traceable directly to the product being manufactured or fabricated, like the fabric in clothing. At Lummus one of the prime costs on a project is the Design Department Man-hours, or Salary.

In a manufacturing operation costs are accumulated through three separate accounts

1. Direct material-cost of materials, assemblies, and parts which are used for the completion of the project.

2. Direct labor-wages of workers who are participating in the completion of the product.

3. Ovehead (also called burden)-the cost of all other factors contributing to the completion of the product.

Material cost consists of the basic purchase price and all other expenses required to transfer the materials to the purchaser's premises, like transportation, insurance, and tariff duties.

Labor cost is made up by many different factors. It includes the basic hourly rate for the hours worked, overtime pay, social security taxes, vacation pay, holidays, sick leave, etc.

INDIRECT COSTS are all the costs of manufacturing that cannot be classified as direct costs, either because it is impractical or impossible. Each classification is initially accumulated in a separate account; and at the end of the accounting period is allocated to individual benefitting activities, like a cost center or project. It is a two-step procedure: accumulation-allocation.

Supervisory services, for example, benefit many units of profit and the cost accumulates in a separate account. At the end of the accounting period the cost is allocated to individual products as part of the overhead.

Modern technology tends to increase the share of indirect costs while the share of direct labor cost is declining. This is really the purpose of automation-to replace direct labor cost with the indirect cost of machines.

All businesses are organized by departments or cost center, but the two are not necessarily synonymous. Costs are accumulated by cost centers, which may or may not coincide with operating departments. Cost centers are located where costs can be measured and recorded as conveniently and accurately as possible. The advantages of having departments and cost centers are

- A more accurate selling price can be gained for a product
- Cost can be controlled more easily
- Long range forecasting can be more exact

The cost engineer should maintain a close liaison with the cost accountants and those who design and apply the cost accumulation system. The cost engineer's requirements must be known in order to be met within the limitation of cost and practicality.

OVERHEAD COSTS. The classification includes all production costs which are not considered prime costs (direct material and direct labor). There is no limit to the number of overhead classifications; however principal groups are

- Indirect materials-also known as supplies, are material like lubricants that do not become a part of the finished product.
- Indirect labor-the wages and salaries of employees who are not directly connected with the manufacture of a product, such as supervisors, maintenance workers, and internal transportation workers. Frequently the cost of fringe benefits is included in this classification.
- Facilities costs-both short-term cost of the current year and the depreciated part of long-term costs for the current year. The former includes building maintenance, local real estate taxes, and others; the latter, buildings, and equipment.
- Service department costs-for facilities which support production, for example accounting, laboratories, stores, cafeteria,

and first aid stations.

Overhead accounting is considered one of the most tedious problems of cost accounting. Accumulated overhead costs are allocated in stages. First, the overhead costs are allocated to the cost centers and in turn are allocated to job order costs and process costs. With a large number of indirect costs classifications and a large number of cost centers, the computations can be voluminous.

A cost engineer must recognize that the cost data that emerge from these calculations are affected significantly by the measure used for allocation. He should understand fully the techniques used for the allocation. A basis must be selected to allocate the costs fairly. Consider, as an example, building maintenance. Measurement can be made on the basis of square feet and allocated on the basis of floor space occupied by each cost center. In turn cost centers must have a measure to allocate this overhead, which might be labor hours, machine hours, material, etc.

A simple illustration will show how overhead is applied to an individual Job at a cost center with allocation based on labor dollars.

Overhead cost for cost center "A"	$50,000
Estimated total direct labor cost	25,000
Burden rate 50,000/25,000 for each direct labor dollar	2
Direct labor cost for Job 500	2,000
Applied overhead 2,000 x 2	4,000

Thus Job 500 would be allocated $4,000 of the total overhead of

$50,000 for cost center A.

Allocating overhead based on actual data has two distinct disadvantages

- It is complicated
- The information is not timely

Units made at the beginning of a month cannot be priced out until the end of the month when actual data become known, and use of actual overhead cost is subject to fluctuation from period to period for items like taxes which may occur only once a year. A short-cut method has been devised called an estimated burden rate. Managers predict an amount for the overhead cost for a fixed period, usually a year, and determine the measure or basis for charging a job or product for its appropriate share of the overhead cost. Direct labor has been used most widely for this basis.

A rate can be established by dividing total estimated overhead cost by the estimated quantity of the selected basis, which becomes the burden rate. These are estimated amounts and variances will occur. The variances should be reviewed periodically and adjusted accordingly.

The advantages of the estimated burden rate method are

- Savings of time
- Increased possibility of obtaining better data for determination of operating efficiency.

STANDARD COSTS. Standard costs and budgets are not identical. Standard costs usually refer to a unit of production, and budget to a total concept like a department. In a sense the standard is the budget for one unit of production. The purposes of standard costs are for

- Building a budget and feedback system
- Aid in management predictions
- Save in bookkeeping cost
- Aid in cash flow forecasting

Standard costs are determined through scientific techniques and objective quantitative measurements. The developed costs do not necessarily represent expected performance but, rather, desired objectives. Standard costs are merely reference costs to which actual costs are compared. The variances are used in management reports as a valuable tool to highlight areas of good and poor performance.

As an example consider a manufacturer producing 10,000 heavy ash trays. Standards are developed for direct materials and direct labor. The standards based on historical data are

4 lb at \$5/lb = \$20/unit	Direct material standard cost
2 hr at \$8/hr = \$16/unit	Direct labor standard cost

On budget of 10,000 units the budget for materials becomes \$200,000 and for labor \$160,000.

After a run of 10,000 units the data compiled are

Good units produced	10,000
Direct material costs	\$270,000
Pound of input	50,000
Material price per pound	\$ 5.40
Direct labor costs	171,600
Hours of input	22,000
Labor price per hour	\$ 7.80

Budget variances are determined as follows with U for unfavorable

	Total Actual costs	Budget total Standard costs	Budget variance
Direct materials	$270,000	$200,000	$70,000 U
Direct labor	171,600	160,000	11,600 U

The variance for materials is $70,000 unfavorable and for labor the variance is $11,600 unfavorable. The variances can be traced in more detail. The $70,000 unfavorable variance for material arose as follows

Difference in quantity 50,000 lb against 40,000 lb	at $5 standard	=	$50,000 U
Difference in price $5.40/lb against $5.00/lb	for 50,000 lb	=	20,000 U
	Total material variance		$70,000 U

The $11,600 unfavorable variance for labor arose as follows

Difference in hours 22,000 hr against 20,000 hr	at $8 standard	=	$16,000 U
Difference in labor rate $7.80/hr against $8 hr	for 22,000 hr	=	4,400 F
	Total labor variance		$11,600 U

Note that the unfavorable total labor variance is the sum of an unfavorable manhour overrun and a favorable labor rate. Without the standard cost, analysis is only meaningful if standards are kept current and accumulation of actual data is kept accurately. Close coordination between accountants and engineers is required.

JOINT COSTS. The term applies to two or more kinds of products that are produced simultaneously and are not identifiable as individual types of products until a certain stage of production or the split-off point is reached. A product for which there is little or no demand is called a by-product. Joint costs are combined cost up to the point of separation. Since joint costs cannot be traced directly to units worked, the apportioning of the costs to various units of production has to be arbitrary. Joint cost distribution is limited to purposes of inventory costing and income measurement.

There are two basic approaches for distributing the cost to various products.

- Physical measures
- Net realizable value (relative sales value method)

Consider the following flow sheet

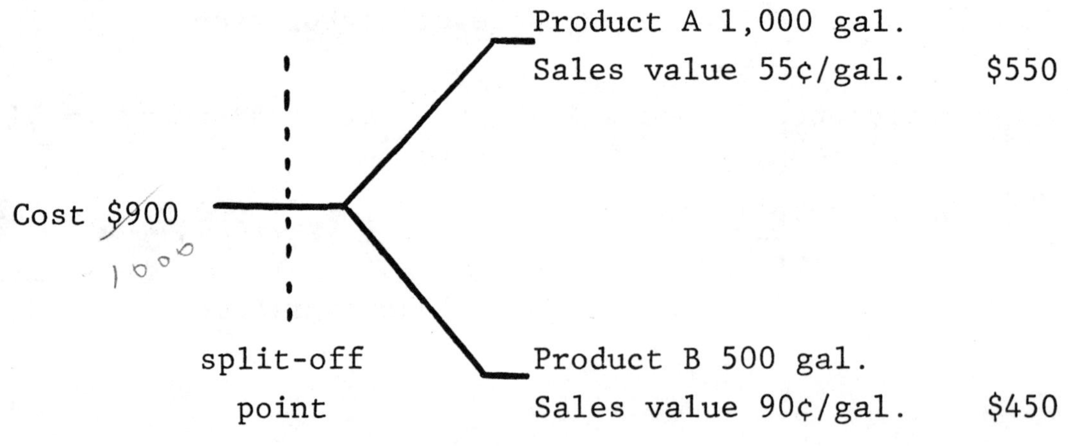

Both of these products become finished goods at the split-off point. Let the $900 cost up to the split-off point be allocated on the basis of the physical measure gallons. For $900 to be spread over 1500 gallon total, the cost per gallon is $0.60 and the allocation becomes

	Production	Cost assigned
A	1,000 gal.	1,000 x 0.60 = $600
B	500 gal.	500 x 0.60 = $300
		$900

The income statement becomes

	A		B		Total
Sales	900 gal	$495	450 gal	$405	$900
Production cost	1,000 gal	$600	500 gal	$300	$900
Less inventory	100 gal	60	50 gal	30	90
Cost of sales	900 gal	$540	450 gal	270	$810
Margin		($45) loss		$135	$90
Margin percentage	$\frac{45}{495}$ = 9.1% loss		$\frac{135}{405}$ = 33.3%		$\frac{90}{900}$ = 10%

There is a loss for "A" and a profit for "B".

Now consider an allocation based on net realizable value

	Net realizable value	Weighting	Cost assigned
A	$550	(550/1000) 900	$495
B	450	(450/1000) 900	405
	$1000		$900

The income statement becomes

	A	B	Total
Sales	$495	$405	$900
Production cost	$495	$405	$900
Less inventory	49.5	40.5	90
Cost of sales	$445.5	$364.5	$810
Margin	$ 49.5	$ 40.5	$ 90
Margin percentage	$\frac{49.5}{495} = 10\%$	$\frac{40.5}{40.5} = 10\%$	$\frac{90}{100} = 10\%$

Now "A" and "B" both show the identical margin percentage of 10%. The cost engineer must be aware of the dangers associated with cost allocations for joint products, since the allocation can have a decisive impact on the cost data.

JOB ORDER AND PROCESS COSTING. The two basic types of accounting techniques for accumulating production costs are

- Job order costing
- Process costing

In job order costing the production cycle and the cost cycle are of equal length. Job order costing is used for specialized production jobs for which costs can be recorded accurately, but considerable work is involved. Job orders can be issued for individual customers or for stock.

Process costing is used for continuous process industries which operate twenty-four hours per day like chemical, steel, and others. The production cycle continues without interruption while the cost cycle is cut off for each accounting period to determine the result of operations. Process costs use averages and are less accurate than job-order costs, but are simpler and cheaper to operate. The method is applicable to mass production industries as will as to process industries.

QUESTIONS

1. Write a brief description of what is meant by the term general and selling expense.

2. What is the basic purpose of the Chart of Accounts?

3. Give examples for several kinds of overhead costs.

4. Discuss the relationship between cost accountants and cost engineers.

5. Write a brief description of what is meant by joint costs.

6. Cost accounting is among the most important parts of (check one)

   Historical financial accounting
   Audit function
   Managerial accounting

7. Give a brief definition and classification of costs.

8. Explain the reasons for the need of allocation of overhead and the techniques used.

9. What are the advantages of having departments and cost centers?

10. Process costing is employed by the following type of industries (check one)

   Consulting engineering
   Chemical industries
   Construction companies

3. COST ESTIMATING

E. E. (Gene) Stackhouse
Petro-Marine Engineering, Inc., Houston, Texas

INTRODUCTION. The purpose of this chapter is to review material for the CCE examination particularly regarding the concepts and practice of cost estimating. Chapters 15 and 16 on Capital Investment Cost Estimation, and Operating Cost Estimation respectively in the book "Cost and Optimization Engineering" are particularly pertinent. Additional information is available in the AACE "Cost Engineers Notebook" and some supplemental references listed at the end of this chapter. In practice a cost engineer should have a personal collection of costs for application to his specialty.

A listing of publications of the AACE is given on the two following pages.

PUBLICATIONS OF THE AMERICAN ASSOCIATION OF COST ENGINEERS

308 MONONGAHELA BUILDING
MORGANTOWN, WEST VIRGINIA 26505

AACE PUBLICATIONS INDEX, 1956-78 (A completely revised and updated keyword-in-context index of all AACE publications and technical papers, including section publications. Includes a bibliography of over 1300 papers on all aspects of cost engineering. Prices are quoted in US $ and are postpaid in U. S. and Canada; **other countries add 50c postage and handling charge.)**

US $15.00 to members
US $25.00 to non-members

ARCHITECTURAL, CONSTRUCTION, MANUFACTURING AND ENGINEERING [ACME] GLOSSARY OF TERMS (An exhaustive glossary of cost engineering and construction terms arranged by CSI divisions and totalling over 500 pages. Prices are quoted in US $ and are postpaid in U. S. and Canada; **other countries add 50c postage and handling charge.)**

US $25.00 to members
US $40.00 to non-members

TRANSACTIONS OF THE FIFTH INTERNATIONAL COST ENGINEERING CONGRESS, 487 pages, 1978 (67 papers on cost engineering topics which were presented at the 5th International Cost Engineering Congress, Utrecht, Netherlands, Oct. 30-Nov. 1, 1978.)

US $50.00 postpaid worldwide

SEMINAR NOTES, FIFTH INTERNATIONAL COST ENGINEERING CONGRESS, 1978 (Detailed notes and outlines of material presented at the feasibility analysis, factored estimating, venture and risk analysis, and project management symposia which preceeded the 5th International Cost Engineering Congress, Utrecht, Netherlands, October 1978.)

US $50.00 postpaid worldwide

(Note: If ordered in conjunction with the **Transactions of the Fifth International Cost Engineering Congress**, the total price of both volumes is US $90.00.)

DIRECTORY OF CERTIFIED COST ENGINEERS (An alphabetical and geographical listing of the names and addresses of all individuals recognized as Certified Cost Engineers as of April 1, 1978.)

US $60.00 postpaid

AACE TRANSACTIONS* (A bound book of technical papers presented at AACE Annual Meetings. Prices listed are quoted in US $ and are postpaid in U. S. and Canada; **other countries add 50c postage and handling charge.)**

1) 1979 Issue US $20.00 each to members
US $30.00 each to non-members
US $22.50 each to NSPE liaison society members
Add US $10.00 to above prices for hard-cover edition

2) 1978 Issue US $20.00 each to members
US $30.00 each to non-members
US $22.50 each to NSPE liaison society members

3) 1977 Issue US $16.00 each to members
US $26.00 each to non-members
US $19.75 each to NSPE liaison society members

4) 1967 Issue (All other issues are out of print)*
US $5.50 each to members
US $10.50 each to non-members
US $8.00 each to NSPE liaison society members

COST ENGINEERING* (A bi-monthly journal; formerly the **AACE Bulletin**)

1) 1-year subscription (calendar year basis only)*
NO CHARGE TO MEMBERS
US $15.00 to non-members (U.S. and Canada)
US $20.00 to non-members (other countries)
US $11.25 to NSPE liaison society members (U.S. and Canada)
US $15.00 to NSPE liaison society members (other countries)

2) Back Issues US $2.00 each to members
US $3.00 each to non-members
US $2.25 each to NSPE liaison society members

*NOTE: All issues of **AACE Transactions** and **Cost Engineering**, including out-of-print issues, are available in microform. To obtain microfilm or microfiche copies write to:

University Microfilms International
Dept. F.A.
300 North Zeeb Road
Ann Arbor, MI 48106 USA

or

University Microfilms International
Dept. F.A.
18 Bedford Row
London, WC1R 4EJ ENGLAND

COST ENGINEERS' NOTEBOOK (A loose-leaf compilation of cost data and information.)

1) Complete set of back issues through 1978 with covers
   US $29.50 to members in U.S. and Canada
   US $35.00 to members in other countries
   NOT AVAILABLE TO NON-MEMBERS

2) Supplementary Issues
   Sent to members at no charge with **Cost Engineering** issues
   (Extra copies, US $1.50 each)
   NOT AVAILABLE TO NON-MEMBERS

3) Extra sets CEN covers, US $2.00 each

COST AND OPTIMIZATION ENGINEERING, F. C. Jelen, ed., McGraw-Hill, New York (A comprehensive text on cost engineering.)

US $23.50 plus 50c postage and handling
(Also available from the publisher and most book stores)

TECHNICAL PAPERS presented at Annual Meetings before 1967. US $1.00 each (specify title and author with order; Xerox copies of out-of-print papers and out-of-print **Transactions** articles are available at a nominal cost.)

SELECTIONS FROM TRANSACTIONS OF THE AMERICAN ASSOCIATION OF COST ENGINEERS, Volume 1, Great Lakes Section, AACE (A compilation of out-of-print articles from 1967-74 **Transactions** which are felt to be of exceptional lasting interest.)

US $4.50 postpaid to U. S. and Canada
US $5.00 postpaid to other countries

SELECTIONS FROM TRANSACTIONS OF THE AMERICAN ASSOCIATION OF COST ENGINEERS, Volume 2, Great Lakes Section, AACE (A compilation of articles from 1975-78 **Transactions** which are felt to be of exceptional lasting interest.)

US $7.50 postpaid to U. S. and Canada
US $8.00 postpaid to other countries

DECISION MAKING: COST, PRODUCTIVITY AND INFLATION (Proceedings of the First Joint AACE/AIIE Conference, Houston, TX, April 1975.)

US $5.00 postpaid to U. S. and Canada
US $5.00 postpaid to other countries

BIBLIOGRAPHY OF INVESTMENT COSTS, OPERATING COSTS AND RELATED INFORMATION

1) 1976 Issue (Contains abstracts of articles on all phases of cost engineering which were published in 25 major journals during 1975.) US $1.20 plus 50c postage and handling

2) 1977 Issue (Contains abstracts of cost engineering articles published in 26 major journals during 1976.) US $2.10 plus 50c postage and handling

1978 TRANSACTIONS OF THE FLORIDA SECTION OF AACE (Contains 13 papers on cost engineering topics which were presented at the 1978 Florida Section Cost Engineering Symposium.)

US $6.00 postpaid in U. S. and Canada
US $6.50 postpaid in other countries

GUIDELINES TO PROFESSIONAL EMPLOYMENT FOR ENGINEERS AND SCIENTISTS. SECOND EDITION.

US $1.00 postpaid

All prices are quoted in US $ and are subject to change without notice. **DO NOT** enclose payment with your order. AACE will invoice you upon shipment.

OTHER PUBLICATIONS AVAILABLE TO AACE MEMBERS

BUILDERS' ESTIMATING FACTBOOK by Roy L. Diez and Lawrence D. Maloney, 1974 Cahners Books. US $10.95.

MANAGING METRICATION IN BUSINESS AND INDUSTRY, Compiled by the American National Metric Council, Marcel Dekker, Inc. US $25.00.

APPLIED COST ENGINEERING by Forrest D. Clark and A. B. Lorenzoni, 1978, Marcel Dekker, Inc. US $24.50.

The above books may be ordered from AACE Headquarters, 308 Monongahela Building, Morgantown, WV 26505. DO NOT SEND PAYMENT WITH YOUR ORDER. The publishers will invoice you directly after shipment.

Orders for these publications must be placed on individual requests or purchase orders for transmittal to the publishers. Orders which include other items cannot be accepted.

FIXED CAPITAL AND WORKING CAPITAL. Capital investment is made up of two components

total investment = fixed capital + working capital

That is, the total investment to put a project in operation is comprised of

- Fixed capital cost--a one-time cost for *all* the facilities for the project and includes land, design and engineering, equipment, utilities, freight, start up, etc.
- Working capital--capital beyond fixed capital needed to meet subsequent obligations. The principal components are inventories and materials, accounts receivable, and cash to conduct the business as for wages and salaries. Generally a one month period can be used. For a rough estimate 10-20% of the fixed capital will serve as an approximation.

TYPES OF ESTIMATES. Some well known types of estimates are

- Current AACE
  - 1. Order of magnitude. -30% to +50%. Use cost capacity curves and cost capacity ratios.
  - 2. Budget. -15% to +30%. Made for the owner's budget. Use flowsheets, layouts, and equipment details.
  - 3. Definitive. -5% to +15%. Use defined engineering data like site data, specifications, basic drawings, and detailed sketches.
- Earlier AACE
  - 1. Order of magnitude (ratio estimate)
  - 2. Study (factored estimate)
  - 3. Preliminary (budget authorization estimate)
  - 4. Definitive (project control estimate)
  - 5. Detailed (firm estimate)
- Edmunds classification
  - 1. Quickie
  - 2. Original
  - 3. Preliminary
  - 4. Official or budget
  - 5. Final or definitive
  - 6. Revised
- Hackney classification
  - 1. Detailed unit cost. ±3%
  - 2. Material takeoff. ±6%
  - 3. Defined equipment ratio. ±12%
  - 4. Preliminary equipment ratio. ±25%
  - 5. Capacity-cost curve. ±40%

COST OF MAKING ESTIMATES. A cost estimate does not come free and the cost rises rapidly with the accuracy required. The following tabulation gives the cost of making an estimate.

Accuracy Range, %	Project cost Mega$ 1	10	20
-5 to +15	$13,000	$35,000	$60,000
-15 to +30	3,200	9,000	14,500
-30 to +50	950	2,500	4,500

FUNCTIONS OF CAPITAL COST ESTIMATES. Capital cost estimates are made for a variety of reasons, the principal ones being

- Feasibility studies
- Selection from among alternative designs
- Selection from among alternative investments
- Appropriation of funds
- Presentation of bids

REPRESENTATION OF COST DATA. Cost estimates can be prepared from three sources of data

- Project estimate - from cost of similar projects, or cost of project components
- Proprietary cost data files - usually from company files
- Published cost information
  - The accuracy is not always known. Sometimes the basis is not clear, for example purchased or installed cost.
  - Extensiveness of cost is sometimes not clear. Some costs are for entire plants while others may be for battery-limits only.
  - Average value is common, but a specific case may be far from average
  - Should be used for order of magnitude estimates

COST INDICES. Costs change continuously because of three factors

- Changing technology
- Changing availability of materials and labor
- Changing value of the monetary unit-inflation.

Various cost indices have been devised to keep up with the changing costs. Cost indices are discussed more fully in Chapter 7. Some frequently used indices are

- Engineering News-Record Index. Construction and building indices prepared for bases of 100 for years 1913, 1920, and 1949.
- Marshall and Swift Equipment Cost Index. It represents the installed equipment cost and is an average for 47 industries. It is a proprietary index derived by a complex procedure involving equipment appraisals, modifying factors, and current economic conditions.
- Nelson Refinery Construction index. Based on 40% material and 60% labor costs.
- Chemical Engineering Plant Cost Index. It is a complex index of 4 major components
    - Equipment machinery and supports 61%
    - Erection and installation labor 22%
    - Buildings, material and labor 7%
    - Engineering and supervision 10%
- Bureau of Labor Statistics. The two indicies used most widely (and in combination) are
    - Labor index
    - Material index
    - (Composite index, like 50:50)

USE AND LIMITATIONS. Cost indices can be used to upgrade a cost for passage of time by a simple proportional relationship. Thus, if a cost was $50,000 in 1958 when an index was 229, the cost in 1965 for an index of 245 would be $53,500

$$50{,}000\frac{245}{229} = \$53{,}500$$

Some limitations of a cost index

- It represents composite data not a single complete project
- The data are averaged
- Varying base periods are used for different indices
- Accuracy may be 10% for a 5 year period, but falls off rapidly for periods greater than 10 years.

COST-CAPACITY FACTORS are used to estimate a cost for a new size or capacity from the known cost for a different size or capacity. The relationship has a simple exponential form

$$C_2 = C_1\left[\frac{Q_2}{Q_1}\right]^X$$

where C_2 = desired cost of capacity Q_2

C_1 = known cost of capacity Q_1

X = cost-capacity factor

If nothing is known use 0.6 as the cost-capacity factor for a single piece of equipment. Thus if a cost for equipment making 100,000 ton/year is 8 Mega$ the cost for equipment making 200,000 ton/year would be 12.1 Mega$.

$$C_2 = 8\left(\frac{200,000}{100,000}\right)^{0.6} = 12.1 \text{ Mega\$}$$

The same procedure can be used for order-of-magnitude estimates for entire plants. If nothing is known use 0.7 as the cost capacity factor for entire plants. Table 3.1 gives some cost-capacity factors for battery-limits plants.

EQUIPMENT INSTALLATION COST RATIOS. A ratio cost factor is merely a factor used to multiply one cost factor by another to get another cost and might be accurate to 20% for the cost of key items. The principle can be used to obtain installed costs from purchased costs. Thus if F denotes an equipment installation cost ratio

Installed cost = (Purchase cost)(F)

An average for F is about 1.5. Some equipment installation factors are given in Table 3.2.

LANG FACTORS. Plant costs can also be approximated from the delivered cost of plant equipment using Lang Factors as multipliers. The multipliers depend on the type of plant and are

- 3.10 for solid process plants
- 3.63 for solid-fluid process plants
- 4.74 for fluid process plants

HAND FACTORS. Instead of using a single multiplier as the Lang Factors are, Hand suggested using a summation of individual fac-

Table 3-1

Cost of Battery-limits Plants*

Process	*Capacity*	*Unit*	*Cost-capacity factor*	*Capacity range*	*Reference, Sec. 15.19*
Acetylene	10	Tons/day	0.73	3.5–250	12
Aluminum (from alumina)	100 M	Metric Tons/yr	0.76	20 M–200 M	17
Ammonia (by steam-methane reforming)	100	Tons/day	0.72	100–3 M	14
Butadiene	10 M	Tons/yr	0.65	5 M–300 M	13
Butyl alcohol	100 MM	Lb/yr	0.55	8.5 MM–700 MM	13
Carbon black	1	Tons/day	0.53	1–150	44
Chlorine	100	Tons/day	0.62	10–800	13, 63
Ethanol, synthetic	10 MM	Gal/yr	0.60	3 MM–200 MM	11
Ethylene	100 M	Tons/yr	0.72	20 M–800 M	45
Hydrogen (from refinery gases)	10 MM	Cu ft/day	0.64	500 M–10 MM	18, 23
Methanol	10 MM	Gal/yr	0.83	5 MM–100 MM	12
Nitric acid (50–60%)	100	Tons/day	0.66	100–1 M	28
Oxygen	100	Tons/day	0.72	1–1.5 M	12, 36, 51, 69
Power plants, coal	100	Mw(elec)	0.88	100–1 M	16
Nuclear	100	Mw(elec)	0.68	100–4 M	16, 60
Styrene	10 M	Tons/yr	0.68	4 M–200 M	11
Sulfuric acid (100%)	100	Tons/day	0.67	100–1 M	6, 13, 28
Urea	250	Tons/day	0.67	100–250	28
Urea	250	Tons/day	0.20	250–500	28

* From O. T. Zimmerman and I. Lavine, *Cost Eng.*, vol. 6, pp. 16–18, July, 1961; and O. T. Zimmerman, *Cost Eng.*, vol. 12, pp. 12–19, October, 1967.

Table 3-2

Typical Equipment Installation Factors

Item	*Installation cost, %*
Belt conveyors	20–25
Bucket elevators	25–40
Centrifugals, disk or bowl	5–6
Top suspended	30–40
Continuous	10–25
Crystallizers	30–50
Dryers, continuous drum	100*
Vacuum rotary	150–200*
Rotary	50–100*
Dust collectors, wet	220–450*
Dry	10–200*
Electrostatic precipitators	33–100*
Electric motors plus controls	60
Filters	25–45
Gas producers	45–250
Instruments	6–300
Ion exchangers	30–275*
Towers	25–50
Turbine generators	10–30

* Includes accessories

Jelen, F. C. Cost and Optimization Engineering
McGraw-Hill Book Co. (1970)

tors multiplied by the delivered cost of different types of equipment. Some Hand Factors are

- 4.0 for fractionating columns, pressure vessels, pumps, and instruments
- 3.5 for heat exchangers
- 2.5 for compressors
- 2.0 for fired heaters

Wroth gave a more complete list as given in Table 3.3.

Table 3-3

Process-plant Cost Ratio from Individual Equipment*

Equipment	*Factor*†
Blender	2.0
Blowers and fans (including motor)	2.5
Centrifuges (process)	2.0
Compressors:	
Centrifugals, motor-driven (less motor)	2.0
Steam turbine (including turbine)	2.0
Reciprocating, steam and gas	2.3
Motor-driven (less motor)	2.3
Ejectors (vacuum units)	2.5
Furnaces (package units)	2.0
Heat exchangers	4.8
Instruments	4.1
Motors, electric	8.5
Pumps:	
Centrifugal, motor-driven (less motor)	7.0
Steam turbine (including turbine)	6.5
Positive displacement (less motor)	5.0
Reactors—factor as approximate equivalent type of equipment	
Refrigeration (package unit)	2.5
Tanks:	
Process	4.1
Storage	3.5
Fabricated and field-erected (50,000 + gal)	2.0
Towers (columns)	4.0

* W. F. Wroth, Factors in Cost Estimation, *Chem. Eng.*, vol. 67, p. 204, October, 1960.

† Multiply purchase cost by factor to obtain installed cost, including cost of site development, buildings, electrical installations, carpentry, painting, contractor's fee and rentals, foundations, structures, piping, installation, engineering, overhead, and supervision.

PLANT COMPONENT RATIOS. Plant cost estimation can be made using factors for separate components, but is generally not more accurate than an overall factor unless a great deal of information is available. Table 3.4 gives a range of plant investment cost for individual components in a battery-limits process plant.

Table 3-4

Process-plant Investment Components

Item	*Percent of total plant investment*
Direct process cost:	
Principal equipment items	15–40
Installation of principal equipment items	6–16
Process piping (installed)	2–25
Instruments (installed)	1–8
Electrical (installed)	1–5
Buildings, process	0.5–15
Average total process costs	53
Other direct costs:	
Utilities (installed)	3–20
General services	0.5–10
Buildings, general	4–14
Receiving, shipping, storage (installed)	2–12
Average total direct costs	78
Indirect costs:	
Engineering, overhead, etc.	8–16
Contingencies	9–22
Average indirect costs	22
Grand total cost	100

Jelen, F. C. Cost and Optimization Engineering
McGraw-Hill Book Co. (1970)

Reprinted with permission

PLANT COST BY ANALYTICAL PROCEDURE. Analytical methods, instead of Tables and Charts, are particulary useful in using a computer for making cost estimates. Three methods suitable for computer calculation are

- Ratio factor method
- Hirsch-Glazier method
- Rudd-Watson Method

The ratio factor method uses a number of factors added together which are multiplied by the total major equipment cost. Thus

$$\left\{ \begin{array}{c} \text{Capital investment} \\ \text{for plant} \end{array} \right\} = (f_1 + f_2 + \ldots)\ \Sigma E$$

where ΣE is the major process equipment cost, and f_1 etc. are cost factors for installation, piping, instrumentation ...

The Hirsch-Glazier method takes the form

$$I = E\,A(1 + F_L + F_P + F_M) + B + C$$

where

- I = total battery-limits investment
- E = indirect cost factor representing contractors' overhead and profit, engineering, supervision, and contingencies. E is normally 1.4
- A = total purchased cost f.o.b. less incremented cost for corrosion-resistant alloys
- B = installed equipment cost
- C = incremented cost of alloys used for corrosion resistance
- F_1 = cost factor for field labor
- F_M = cost factor for miscellaneous items
- F_P = cost factor for piping materials

The factors F_L, F_M, and F_P are not simple ratios but are defined by logarithmic relationships. For more details see Cost and Optimization Engineering Chap. 15 or the original reference Chem. Eng. Progr. vol 60, pp 23-25, December, 1964.

The Rudd-Watson method takes the form

$$C_{FC} = \phi_1 \phi_2 \phi_3\ \Sigma C_{EQ}$$

where

- C_{FC} = fixed capital investment
- ϕ = components for installation etc.
- C_{EQ} = purchased equipment cost

Note that in the Rudd-Watson method the factors are multiplied together where as in the ratio factor method they are added together.

Table 3-5

Typical Cost Analysis of Building for Printing Plant†

Frame	Reinforced concrete
Exterior walls	Masonry
No. of floors, excluding basement	10
No. of basement floors	1
Basement area, sq ft	21,000
Gross area supported (excluding slab on grade), sq ft	229,000
Face brick area, sq ft	40,500
Curtain walls, including glass, sq ft	30,000
Net finished area, sq ft	208,000
No. of elevators	3
Story height, typical floor, ft	12

Item	*Unit*	*No. of units*	*$ per unit*	*Cost, $M*	*Cost, %*
General conditions and fee	Sq ft	229,000	1.48	340	9.00
Excavation	Sq ft	21,000	6.48	136	3.60
Foundation	Sq ft	21,000	10.81	227	6.01
Concrete, formed	Sq ft	229,000	5.79	1,326	35.12
Exterior masonry	Sq ft	40,500	7.31	297	7.87
Interior masonry				83	2.20
Stone, granite, marble				44	1.17
Structural steel	Sq ft	229,000	1.32	302	8.00
Miscellaneous metal, including stairs	Floor	11	6,182	68	1.80
Carpentry				41	1.09
Waterproofing and dampproofing	Sq ft	30,000	0.20	6	0.16
Roofing and flashing	Sq ft	21,000	1.38	29	0.77
Metal doors and frames	Floor	10	9.00	9	0.24
Metal windows	Sq ft	30,000	1.13	34	0.90
Hardware	Floor	10	60	6	0.16
Glass and glazing	Sq ft	30,000	0.90	27	0.72
Curtain walls	Sq ft	30,000	0.93	28	0.74
Lath and plastic	Sq ft	208,000	0.02	5	0.13
Tile work				12	0.32
Acoustical ceiling				3	0.08
Toilet partitions				2	0.05
Elevators	Each	3	49,667	149	3.95
Plumbing				120	3.18
Heating and ventilation	Sq ft	208,000	0.85	176	4.66
Electrical	Sq ft	208,000	1.42	296	7.84
Miscellaneous trades	Sq ft	208,000	0.05	10	0.26
Total				3,776	100

* From *Eng. News-Record*, vol. 180, p. 139, June 20, 1968.
† Building in Brooklyn, N.Y.; costs as of June, 1968.

BUILDING COSTS can be estimated from

- ·Unit costs, like $/square foot
- ·Component costs, like floor, wall, roof etc.
- ·Knox volumetric ratio concept-process equipment occupy 3.75% of building volume

Table 3-5 gives some building costs.

ENGINEERING COSTS are approximately 10% of a project cost but vary with the size and type of job. Fig. 3-1 gives a correlation.

Fig. 3-1

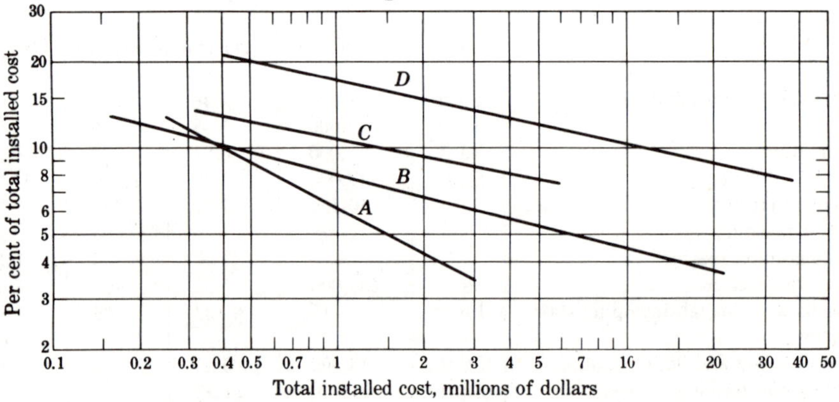

Engineering costs as average percent of total installed costs. *A*, office buildings and laboratories; *B*, power plants, cement plants, kilns, and water systems; *C*, battery-limits chemical process plants; *D*, complex chemical and grass-roots chemical plants and pilot plants.

Reprinted with permission from Jelen, F.C. Cost and Optimization Engineering McGraw-Hill Book Co. (1970)

PIPING COSTS vary from 1% to 25% of project cost. Installation costs are 3 to 4 times purchase costs for common materials. Order of magnitude estimates can be obtained by factors or ratios. Detailed estimates involve quantity takeoffs, labor, and costing.

Intermediate estimates utilize tables and charts for average piping components. The *diameter-inch method* is based on

$$\binom{\text{number of}}{\text{connections}} \times \binom{\text{pipe}}{\text{diameter}} \times \binom{\text{labor}}{\text{factor}}$$

An average is 1.5 man-hours per diameter-inch.

Ratios are available for materials other than carbon steel.

Several manuals are available for estimating piping costs. Figs. 3-2 and Table 3-6 give some compact information on the two components of piping costs--materials, and installation labor.

The two components for piping costs are

- Materials
- Installation labor

The cost for piping materials like straight pipe fittings and valves is generally well established for a host of materials ranging from black iron to stainless steel to plastics. Most companies keep accurate information on the man-hours of labor required to install pipe and fittings for the various sizes and materials in use.

INSULATION COSTS usually run 1% to 2% of plant costs.

START UP COSTS are the costs of transitioning from construction to operation. They include costs to make changes so the plant can operate at design or near design level. Start up costs are generally about 10% of the project cost.

A WORKSHEET FOR COST ESTIMATES is given in Table 3.7.

CURRENT TRENDS IN CAPITAL COST ESTIMATING. Projects are becoming very large in scale and competition is severe and world wide. More factors are being considered like

- Technology
- Inflation
- Price controls
- Regulations on safety, pollution, social aspects

Fig. 3-2

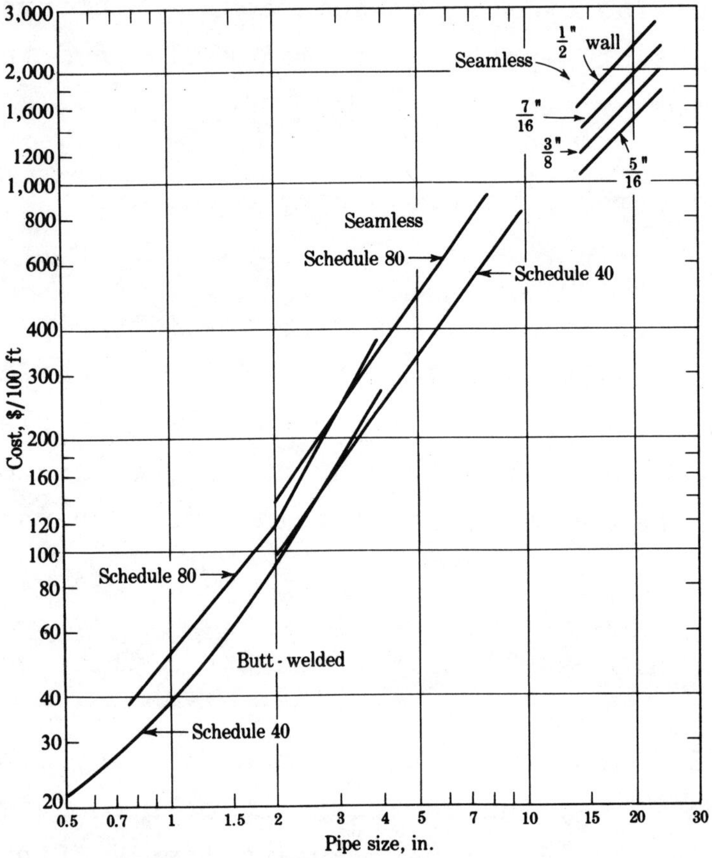

Approximate delivered cost of wrought steel pipe (1978). Schedule 40 and 80 are nominal pipe size; large pipe is outside diameter. Curves are for black plain-end pipe in standard lengths.
For galvanized pipe add 20%;
for threads and couplings add:
schedule 40, black 6%, galvanized 5%;
schedule 80, black 11%, galvanized 10%.

Based upon Jelen, F.C. Cost and Optimization Engineering McGraw-Hill Book Co. (1970) by permission

Table 3-6

Installation Labor Requirements for Carbon Steel Piping

(1)	(2)	(3)	(4)	(5)	(6)	(7)	(8)	(9)	(10)	(11)
Nominal pipe size, in.	*Butt welds per connection*		*Flange joint handling per flange‡*		*Flanged valve handling per value§*		*Pipe handling, per foot of pipe¶*		*Screwed joints per connection***	
	*Std. wt.**	*Extra heavy†*	150 *lb*	300 *lb*	150 *lb*	300 *lb*	*Std wt.*	*Extra heavy*	*Malleable iron*	*Forged steel*
¾	0.80	0.90	0.6	0.7	0.3		0.7	0.08	0.4	0.6
1	0.95	1.1	0.7	0.8	0.4		0.08	0.09	0.5	0.8
1¼	1.1	1.3	0.7	0.8	0.5		0.09	0.10	0.6	1.0
1½	1.2	1.4	0.8	0.9	0.6	0.7	0.10	0.11	0.8	1.4
2	1.4	1.7	0.8	1.1	0.7	0.8	0.11	0.14	1.2	2.0
2½	1.8	2.1	0.8	1.4	0.8	1.0	0.14	0.16	1.5	
3	2.0	2.5	0.9	1.4	1.0	1.2	0.16	0.20		
4	2.5	3.0	1.1	1.4	1.3	1.5	0.20	0.25		
6	3.4	4.5	1.3	2.0	1.7	2.6	0.30	0.40		
8	4.0	6.2	1.3	2.5	2.5	3.8	0.40	0.50		
10	5.7	7.4	2.2	3.5	3.5	5.2	0.50	0.65		
12	6.0	8.5	2.2	4.0	4.5	7.5	0.60	0.70		
14	7.0	9.4	2.6	5.0	6.0	10.0	0.65	0.78		
16	7.9	10.5	3.5	6.0	7.0	13.0	0.70	0.85		
18	8.8	11.9	4.0	6.5	10.0	16.0	0.75	0.90		
20	9.7	13.1	4.5	7.0	12.0	20.0	0.80	1.00		
24	11.4	15.8	5.5	9.0	17.0	30.0	0.95	1.15		

* Man-hours per connection to receive, store, handle through fabrication, fit, weld, and align on pipe racks; includes standard-weight tees, elbows, reducers, and 150-lb standard weld-neck flanges. For 5% chrome, stainless steel, brass, aluminum, and Monel piping add 100%.
† Same as (2) except extra heavy fittings and 300-lb standard weld-neck flanges.
‡ Man-hours for installing gasket and bolting tight.
§ Man-hours to receive, store, erect, and align on pipe racks, ready for bolting up.
¶ Man-hour to receive, store, handle through fabrication, erect, and align on pipe racks ready for bolting or welding.
** Man-hour to receive, store, handle, cut, thread, fit, and make up service-tight; includes tees, elbows, unions, reducers, couplings, and valves.

Jelen, F. C. Cost and Optimization Engineering McGraw-Hill Book Co. (1970)

Table 3-7

COST ESTIMATE

AC NO.	ITEM & DESCRIPTION	MANHOURS	ESTIMATED COST			
			LABOR	SUB-CONTRACTS	MATERIALS	TOTAL
00.00	Excavation					
10.00	Concrete					
20.00	Structural Steel					
30.00	Buildings					
40.00	Machinery & Equipment					
50.00	Piping					
60.00	Electrical					
70.00	Instruments					
80.00	Painting & Scaffolding					
85.00	Insulation					
	DIRECT FIELD COSTS					
90.00	International Expense					
91.00	Temporary Construction Facilities					
92.00	Constr. Services, Supplies & Expense					
93.00	Field Staff, subsistence & Expense					
94.00	Craft Benefits, Payroll Burdens & Insur.					
95.10	Equipment Rental					
95.50	Small Tools					
99.40	Field Staff Overhead Costs					
	INDIRECT FIELD COSTS					
	TOTAL FIELD COSTS					
96.00	Home Office Construction					
	Project Engineering					
	Process Engineering					
	Design					
	Purchasing					
	Business Services					
97.00	Office Expense					
98.00	Office Payroll Burdens					
99.50	Office Overhead Costs					
	TOTAL OFFICE COSTS					
	TOTAL FIELD & OFFICE COSTS					
99.60	Fee					
99.30	Sales Tax					
99.10	Escalation					
99.20	Contingency					
	TOTAL					

Some trends are

- ·More data tracking
- ·More frequent updating
- ·More cost estimating, evaluations, predictions, and control
- ·More computer aided routines
- ·More recognition of organizations, publications, and education

CONSTRUCTION COSTS. Numerous articles and publications are available and are shown in principle in the following pages some of which are cut up assemblies. They are in order

·Some costs for site development and some unit costs for site development--two pages.

·Guide to Information Sources in the Construction Industry. The book lists sources for information exclusively--two pages.

·Design Cost File. This guide to construction costs gives details for estimating a wide variety of structural components like foundations, walls, roofing, and electrical systems among others.

·Building Cost Manual. Has current construction or replacement costs for residential, commercial, industrial, agricultural, and military buildings.

·The Building Estimator's Reference Book. Contains detailed information on site work, concrete, carpentry, mechanical, electrical, etc. A condensed Vest Pocket Estimator is included.

·Building construction books. Six books covering topics like building system cost guide, building cost data, project management, labor rates, and practical critical path method.

·Richardson Rapid Construction Cost Estimating System - the Richardson System. A detailed series of volumes.

·Gulf Publishing Company Manuals. Six manuals by John S. Page covering man-hour requirements, and equipment installation costs.

·American Association of Cost Engineers. Cost Engineers' Notebook. The Table of Contents is reproduced--three pages.

1969 Costs for site development

		Field Installation, $ *M&L*			*L/M*
	Unit	Min.	Norm.	Max.	Ratio
Dewatering and drainage					
Pumping system (rented)	Day	25	32	40	.....
Wellpoint dewatering system	Month	6,500	7,500	8,500	.....
Drainage trench	Lin. ft.	0.75	0.85	0.95	.....
Fencing					
Complete fence (light)	Lin. ft.	1.34	1.88	2.42	0.32
Complete fence (heavy)	Lin. ft.	1.51	2.13	2.75	0.32
Chain link	Lin. ft.	5.48	5.93	6.38	0.48
Gates 6 ft. (light)	Ea.	55.50	67.50	79.50	0.15
(heavy)	Ea.	69.25	82.50	95.75	0.15
(chain)	Ea.	105.65	128.00	150.65	0.15
Corner posts	Ea.	31.50	32.00	32.50	0.19
Fire protection					
Pumps Firehouse Firetrucks (2)	Allowance	100,000	150,000	200,000	.....
Land surveys and fees					
General surveys and fees	% total cost	4.0	9.0	14.0	.....
Soil tests	Ea.	300	400	500	.....
Landscaping					
General	Sq. yd.	1.50	1.70	1.90	0.90
Piling					
Wood (untreated)	Lin. ft.	1.70	2.15	2.60	2.20
Wood (creosoted)	Lin. ft.	2.15	2.60	2.90	2.50
Concrete: precast	Lin. ft.	6.75	7.00	7.25	0.50
cast in place	Lin. ft.	4.75	6.62	8.50	1.25
Steel pipe (concrete filled)	Lin. ft.	7.50	9.50	11.50	0.25
Steel section	Lin. ft.	7.40	8.50	9.50	0.27
Sheet piling, steel	Sq. ft.	1.45	2.60	3.75	0.20
wood	Sq. ft.	1.25	1.75	2.25	0.35
Pile driver setup	Ea.	6,800	7,500	8,200	.....
Roads, walkways, paving					
aving					
4-in.-thk. reinf., 6-in. subbase	Sq. yd.	6.35	7.87	8.39	1.75
6-in.-thk. reinf., 6-in. subbase	Sq. yd.	7.61	9.37	10.13	1.75
2-in.-thk. asphalt top, existing base	Sq. yd.	2.37	3.12	3.87	0.22
2-in.-thk. asphalt top, 4-in. subbase	Sq. yd.	3.58	4.68	5.78	0.22
3-in.-thk. asphalt top, 12-in. subbase	Sq. yd.	6.37	7.62	8.87	0.22
Gravel surface					
2-in.-thk. gravel	Sq. yd.	0.33	0.58	0.83	0.52
4-in.-thk. gravel	Sq. yd.	0.55	0.87	1.19	0.52
6-in.-thk. gravel	Sq. yd.	1.00	1.38	1.76	0.52
Parking lots					
Black-top surface	Sq. yd.	5.30	6.25	7.54	0.45
Sewer facilities					
Asbestos Cement pipe (general)	Lin. ft.	4.55	4.85	5.15	0.38
Concrete pipe (reinforced)					
18 in. dia	Lin. ft.	5.65	5.80	5.95	0.38
36 in.	Lin. ft.	14.75	15.96	17.23	0.39
72 in.	Lin. ft.	50.33	52.18	54.03	0.40
Vitrified clay piping					
18 in. dia	Lin. ft.	7.55	7.80	8.45	0.86
24 in.	Lin. ft.	15.15	16.10	17.05	0.88
36 in.	Lin. ft.	33.95	36.20	38.95	0.89
Septic tank (45,000 gal.)	Ea.	......	7,500	......	0.05
Site clearing, excavation and grading					
Site preparation					
Machine cuts	Cu. yd.	0.50	0.56	0.63	0.30
Clearing and grubbing	Sq. yd.	0.13	0.15	0.18	.....
General grading	Sq. yd.	0.63	0.44	0.48	.....
Final leveling	Sq. yd.	0.25	0.31	0.38	.....
Foundation excavation					
Machine excavation	Cu. yd.	1.50	1.63	1.75	0.58
Machine plus hand trim	Cu. yd.	2.50	3.44	3.75	0.90
Hand work	Cu. Yd.	7.56	10.00	12.50	.....
Trench excavation					
Machine 3½ ft. deep × 2 ft. wide	Lin. ft.	0.38	0.44	0.50	0.50
4 ft. × 3 ft.	Lin. ft.	0.56	0.63	0.68	0.50
Machine 4½ × 4 ft.	Lin. ft.	1.12	1.13	1.25	0.50
Machine 5 ft. × 5 ft.	Lin. ft.	1.38	1.50	1.63	0.50
Hand labor	Cu. yd.	8.75	10.12	13.75	.....
Trench shoring					
Sheeting	Sq. ft.	1.25	1.52	1.75	1.12
Trench and foundation backfill					
Machine plus hand trim	Cu. yd.	1.44	1.56	1.68	0.90
Hand labor only	Cu. yd.	5.79	6.25	6.75	.....
Miscellaneous materials					
Sand	Cu. yd.	3.05	4.80	5.55	.....
Gravel	Cu. yd.	1.50	2.25	3.00	.....
Dirt fill	Cul yd.	1.30	2.15	3.00	.....
Crushed stone	Cu. yd.	2.55	4.37	5.19	.....

K. M. Guthrie, Author

1969 Unit costs for offsite facilities

	Unit	Field Installation, $ — M&L Min.	Norm.	Max.	L/M Ratio
Air systems					
Instrument air					
Compression facilities, air dryer, air receiver, and distribution	$M	18.75	43.75	62.50	0.80
Plant air					
Compression facilities, air receiver and distribution	$M	12.50	31.25	50.85	0.80
Blowdown and flare					
For general purposes (including flare lines, blowdown drum and disposal pit)	$M	81.75	102.58	187.52	0.45
Cooling tower and CW distribution					
Use 1.15 design factor on estimated throughput.					
Cooling tower costs		See cost data section			
Distribution systems					
For general purposes	Gpm.	12.58	36.50	43.25	0.85
River intake installation					
For general purposes	Gpm.	8.22	16.25	24.37	1.12
Fireloop and hydrants					
For general purposes	$M	12.58	22.50	40.24	0.80
Fuel systems					
Fuel oil (includes pumps, storage, piping, controls and distribution)	$M	6.25	25.12	43.75	0.85
Fuel gas (includes receiver, piping, controls and distribution)	$M	12.50	37.52	62.58	0.85
General water systems					
Treated water					
Filtered and softened	Gal.	0.15	0.23	0.30	
Distilled	Gal.	0.65	0.92	1.20	
Drinking & service water					
General facilities	$M	2.50	5.40	7.58	
Power generation and distribution					
Use 1.10 design factor on estimated consumption.					
Generating facilities	Kw.	See cost data Section			
Electrical distribution					
For general purposes	Kw.	87.5	93.75	98.75	0.75
Main transformer stations					
Three phase, 60 cycle					
Capacity 3,000 kva.	Kva.	33.0	37.0	44.0	
5,000	Kva.	20.0	23.0	26.0	
10,000	Kva.	13.0	14.0	16.0	
20,000	Kva.	10.0	12.0	13.0	
Secondary transformer stations					
4,200/575 v.					
600 kva	Kva.	30.1	33.8	42.3	
1,000	Kva.	20.1	25.2	31.5	
1,500	Kva.	15.6	19.5	24.3	
2,000	Kva.	14.8	18.5	23.2	
13,200/575 v.					
600 kva	Kva.	28.2	35.3	44.2	
1,000	Kva.	21.2	26.5	33.1	
1,500	Kva.	16.6	20.8	26.2	
2,000	Kva.	15.4	19.3	24.1	
Receiving, shipping, storage					
Automotive					
Forklift trucks					
3,000 lb.	Ea.		7,800		
5,000 lb.	Ea.		11,000		
10,000 lb.	Ea.		16,200		

	Unit	Field Installation, $ — M&L Min.	Norm.	Max.	L/M Ratio
Pallet truck					
Hydraulic 4,000 lb.	Ea.		930		
Electric 4,000 lb.	Ea.		3,600		
Payloaders					
2 cu. yd. (gas)	Ea.		21,000		
4 cu. yd. (gas)	Ea.		33,700		
2 cu. yd. (diesel)	Ea.		22,900		
4 cu. yd. (diesel)	Ea.		36,500		
Tank trailers					
Carbon steel	Ea.		14,800		
Aluminum	Ea.		21,600		
Stainless	Ea.		36,500		
Tractors					
Gasoline	Ea.		12,500		
Diesel	Ea.		27,500		
Tractor shovel					
2-cu. yd. bucket	Ea.		26,300		
3-cu. yd. bucket	Ea.		28,700		
4-cu. yd. bucket	Ea.		35,600		
Automotive shipping facilities					
One outlet per 2,000 bbl.			9,800		0.85
Docks and wharves					
Light construction					
2-in. deck	Sq. ft.	5.15	5.63	6.25	0.45
3-in.	Sq. ft.	6.25	6.87	7.50	0.45
Medium construction					
3-in.	Sq. ft.	8.75	9.38	10.12	0.45
4-in.	Sq. ft.	10.15	11.25	12.50	0.45
Heavy construction					
4-in.	Sq. ft.	12.50	15.62	18.75	0.45
concrete	Sq. ft.	17.50	21.25	25.25	0.45
Dredging					
General operations	Cu. yd.	4.32	10.81	17.28	
Tankage					
General		See cost data section			
Railroad					
Straight track (railroad siding)	Lin. ft.		26.25		0.58
Turnout	Ea.		2,800		0.10
Bumper	Ea.		790		0.12
Blinker and gate	Ea.		9,300		0.15
Grade and ballast	Lin. ft.		6.25		0.95
Locomotives (battery)					
9 ton	Ea.		35,000		
12 ton	Ea.		41,800		
Locomotives (diesel)					
1½ tons	Ea.		11,000		
3 tons	Ea.		14,000		
Tank car (10,000 gal.)	Ea.		10,800		
Railroad shipping facilities					
One outlet per 2,000 bbl.	Ea.		4,800		0.85
Steam generation and distribution					
Use 1.10 design factor on estimated consumption.					
Package boilers (up to 150,000 lb./hr.)	Lb./hr.	See cost data section			
Field erected (above 150,000 lb./hr.)	Lb./hr.	See cost data section			
Steam distribution					
For general purposes	Lb./hr.	0.94	1.52	1.68	0.85
Yard lighting and communications					
For general purposes	$M	18.75	52.25	93.75	0.75
Yard transfer lines and pumps					
For general purposes	$M	17.25	31.25	56.25	0.65

K. M. Guthrie, Author

GUIDE TO INFORMATION SOURCES IN THE CONSTRUCTION INDUSTRY

Compiled by
JULES B. GODEL

Construction Publishing Company Inc.

CONTENTS

Reprinted by permission from Godel, J.B. Guide to Information Sources in the Construction Industry Construction Publishing Co., Inc. (1975)

#5 EXTERIOR WALLS
UCI #04210

BRICK & BLOCK CAVITY WALL

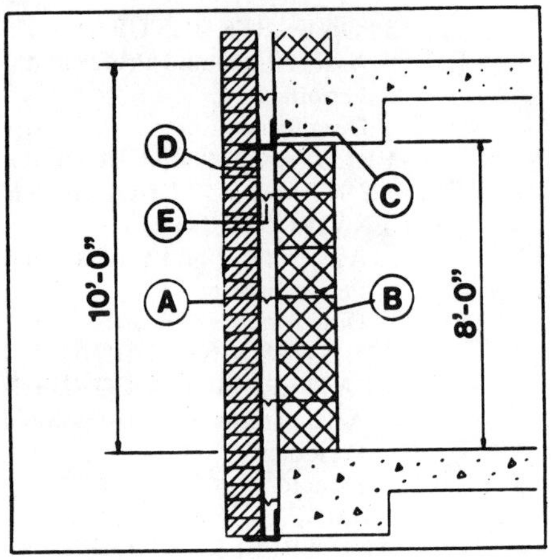

GENERAL INFORMATION

A - 4" FACE BRICK

B - CONCRETE BLOCK

C - STEEL LINTLE

D - BACK PARAGE

E - WALL TIES & ACCESS

COMMON BOND

SERIES	WALL THICKNESS	DESIGN FACTORS				DIRECT COSTS	
		F.R.	STC	"U"	DL	SF	LF
13024	10 INCH	4 HR	82	0.17	77#/SF	$7.23	$72.30
13025	12 INCH	4 HR	89	0.14	80#/SF	$7.38	$73.80
13026	14 INCH	4 HR	95	0.13	86#/SF	$7.68	$76.80

TABLE OF CONTENTS

Construction Publishing Company Inc.
Compiled by:
McKee-Berger-Mansueto Inc.

Looking For the Cost or Value of Buildings in 1980?

Structures Cost Manual has current construction or replacement costs for 47 types of residential, commercial, industrial, agricultural and military buildings. In a few minutes you work up a reliable budget estimate based on the actual materials and design features, area, shape, wall height, number of floors and support requirements. Most important, *Structures Cost Manual* covers all of the important variables that can make any building unique from a cost standpoint.

Nearly all the figures you need for each building are brought together on a single page. You get a reliable total cost for the building you have in mind and for the location you select. 248 pages 8½ x 11 $10.00

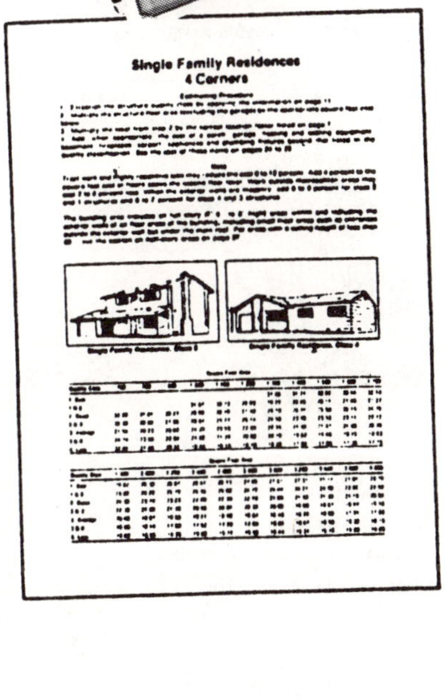

If you need square foot building costs or replacement values in 1980 this will be the most valuable reference you own.

ADJUSTERS	Determine the current replacement cost for your area and the actual cash value. Establish the proper recovery. Verify insurance to value.
APPRAISERS	Verify your calculations, develop values for unusual building types, add or delete the value of less common building features.
ARCHITECTS	Compare the large number of figures for each variable in this book with your plots of value trends for similar variables.
CONTRACTORS	Prepare budget estimates; check your own bid figures.
INSURANCE OFFICERS	Establish current replacement costs; demonstrate the dangers of under-insurance.
LOAN OFFICERS	Check the accuracy of construction estimates; establish the full value of all proposed improvements before loan commitment.

CRAFTSMAN
Book Company
of America
542 Stevens Ave
Solana Beach,
California 92075

Reprinted with permission

Reprinted with permission

R.S. MEANS CO., INC.
1979 CATALOG — A

BUILDING CONSTRUCTION BOOKS
SEMINARS
ESTIMATING FORMS

Estimating Forms,
Seminars and Books

R.S. Means Co., Inc.

Duxbury, MA 02332 (617) 934-6571

GENERAL NOTES AND INSTRUCTIONS PAGE 1

ALL DATA CONTAINED IN THE RICHARDSON PROCESS PLANT CONSTRUCTION ESTIMATING STANDARDS IS ARRANGED BY ACCOUNTS AS FOLLOWS:

EDITED & PUBLISHED BY
RICHARDSON ENGINEERING SERVICES, INC.
POST OFFICE BOX Y
SOLANA BEACH, CALIFORNIA 92075
714: 755-5181

PREPARED BY
INTERNATIONAL CONSTRUCTION ANALYSTS
POST OFFICE BOX 370
SOLANA BEACH, CALIFORNIA 92075
714: 755-6776

THIS IS THE 1978-1979 EDITION OF:

THE RICHARDSON RAPID CONSTRUCTION COST ESTIMATING SYSTEM

Cost Estimating Manual for Pipelines and Marine Structures

John S. Page

Gulf Publishing Company
Book Division
Houston, Texas

Man-Hour Manuals by John S. Page

Estimator's Equipment Installation Man-Hour Manual

Estimator's General Construction Man-Hour Manual

Estimator's Man-Hour Manual on Heating, Plumbing, Air Conditioning, and Ventilating

Estimator's Piping Man-Hour Manual (co-author)

Estimator's Electrical Man-Hour Manual (co-author)

Other Books by John S. Page

Estimator's Manual of Equipment and Installation Costs

TABLE OF CONTENTS

TABLE OF CONTENTS, continued

AMERICAN ASSOCIATION OF COST ENGINEERS	INDEX NUMBER AA-0.000	3 of 3
COST ENGINEERS' NOTEBOOK	REVISION 4	DATED 1/79
	REPLACES 3	8/77

TABLE OF CONTENTS, continued

* Denotes items issued 1970 or earlier. Index numbers for these items are subject to change upon revision.

Chapter 3. Part 2. Operating-Cost Estimation

DEFINITIONS

- Operating cost or manufacturing cost is the expense in keeping a project, operation, or piece of equipment running and producing.
- General expense is the administrative, sales, and general expenses associated with operating cost.
- Total product cost is the sum of operating cost (manufacturing cost) and general expense.
- Direct costs, also called variable costs, are operating costs directly apportioned to throughput.
- Indirect costs, or fixed costs, are operating costs that are independent of throughput.

Operating-cost estimation--general

- Purposes of operating-cost estimates
  - Screen alternative projects
  - Guide research efforts
  - Evaluate component sensitivity
- Important factors
  - Calculate costs at full and reduced levels of production
  - The cost may be calculated as it stands alone, or as an incremented cost for a specific project.
- Three common bases for operating costs
  - Daily
  - Unit-of-product
  - Annual
- Advantages of annual basis
  - Evens out seasonable variations
  - Considers on-stream time factor
  - Adapts better to less than full capacity operation
  - Considers readily infrequently occurring large expenses
  - Gives calculations more directly usable in profitability analysis

RAW MATERIALS

- Raw materials frequently largest operating cost
- By-products and scrap may be a credit
- Prices obtained from sources or published data
- Catalysts and supplies are considered raw materials
- Fuels may be raw material or utility
- Freight and handling costs must be included

UTILITIES. Table 3-8 gives some rates for industrial utilities.

Table 3-8. Typical Rates for Industrial Utilities

Utility	1979 Cost, $	Unit
Steam:	2.80-4.80	1,000 lb
Electricity:		
Purchased	0.025-0.040	Kwhr*
Self-generated	0.020-0.065	Kwhr*
Natural Gas:	2.00-3.00	1000SCF
Fuel Oil:	0.30-0.90	Gallon
Coal:	25.00-40.00	Ton
Water:		
City	0.30-1.50	1000 gal
Filtered and softened	0.40-1.00	1000 gal
Distilled	2.00-3.50	1000 gal

*Highly dependent upon load factor and location

OPERATING LABOR. Best estimates are based on a complete manning table.

The Wessel equation is

$$\frac{\text{operating man-hr}}{\text{tons of product}} = T \frac{\text{number of process steps}}{(\text{capacity, ton/day})^{0.76}}$$

T = 23 batch operations with maximum labor
T = 17 for operations with average labor requirements
T = 10 for well instrumented continuous process operations

Supervision is typically 20% of operating labor.

MAINTENANCE

- Little published data
- Estimated at 3% to 10% of project cost per year
- Generally considered to be 60% labor 40% material
- Maintenance varies with rate of operation
  - 85% at 75% of capacity
  - 75% at 50% of capacity
- Maintenance increases with age, but average values commonly used

INDIRECT PAYROLL COST includes workmen's compensation, pensions, group insurance, paid vacations and holidays, social security, unemployment taxes, fringe benefits etc. It is generally based on labor cost, usually at 20% to 30% of labor cost; but it is increasing with time.

OPERATING SUPPLIES include lubricants, instrument charts, brooms etc. and may be assumed to be 6% of operating labor.

LABORATORY AND OTHER SERVICE COSTS can be based on

- Past experience
- Work, like man-hr, required
- A percentage of operating labor cost-3% to 10% but may be as high as 20%

ROYALTIES AND RENTALS are generally an operating expense but may be part of capital investment. Royalty payments may range between 1% to 5% of product sales price.

CONTINGENCIES. Direct cost contingencies range from 1% to 5% of direct cost.

INDIRECT COSTS. Factory overhead, indirect cost, has two groupings

- One dependent on investments
- One dependent on labor

Some data on the allocation for new plants follows. The figures represent what percentage of the investment and labor respectively can be used for estimating the factory overhead for the two classifications

	Investment factor, % per year	Labor factor, % per year
Heavy chemical plants, large capacity	1.5	45
Power plants	1.8	75
Electro chemical plants	2.5	45
Cement plants	3.0	50
Heavy chemical plants, small capacity	4.0	45

Factory overhead is also known as general <u>works</u> expense, but not general expense.

DISTRIBUTION COSTS include handling and transportation costs. The costs vary with the types of containers and methods of shipment. Cost without distribution cost is called bulk cost.

AVOIDANCE OF NUISANCES includes waste disposal and pollution control costs. Each case must be calculated individually. The topic is becoming increasingly more extensive including such items as product liability. The cost is mounting rapidly.

SHORTCUT METHODS. Table 3-9 will help in making preliminary operating-cost estimates.

Table 3-9

Preliminary Operating-cost Estimate

A. Direct production costs
  1. Materials
    a. Raw materials—estimate from price lists
    b. By-product and scrap credit—estimate from price lists
  2. Utilities—from literature or from similar operations
  3. Labor—from literature or from similar operations
  4. Supervision—10 to 25% of labor
  5. Payroll charges—15 to 25% of labor plus supervision
  6. Maintenance—3 to 6% of investment per year
  7. Operating supplies—0.5 to 1.0% of investment per year
  8. Laboratory—20% of labor
  9. Royalties—1 to 5% of sales
  10. Contingencies—2 to 10% of direct costs

B. Indirect costs
  1. Depreciation—5 to 10% of investment per year
  2. Property taxes—2% of investment
  3. Interest—6 to 8% of investment
  4. Insurance—1% of investment
  5. Plant overhead—40 to 60% of labor or 15 to 30 percent of direct costs

C. Distribution costs
  1. Packaging—estimated from container costs
  2. Labor—from similar operations
  3. Shipping—from carriers or 1 to 3% of sales
  4. Overhead—50 to 75% of distribution labor cost

Jelen, F. C. Cost and Optimization Engineering McGraw-Hill Book Co. (1970)

Reprinted with permission

COMPONENTS OF TOTAL PRODUCT COST. Table 3-10 will serve to illustrate the connection between the component cost items that enter the total product cost. There is variation from company to company but the tabulation will be a good guide.

SUMMARY. The basic cost estimate procedure is

·Scope the job or problem	Understand the *overall* viewpoint
·Establish estimate formate	Code of accounts
·Prepare estimate	Standard estimate
Area by area	Document well
Category by category	
·Check Work. Verify data	Analyze
·Review and adjust	Tailor to job
·Finalize.	Plan ahead
Complete documentation	
File	

Table 3-10

Components of Total Product Cost*

- I. Operating cost or manufacturing cost
  - *A*. Direct production costs
    - 1. Materials
      - *a*. Raw materials
      - *b*. Processing materials
      - *c*. By-product and scrap credit
      - *d*. Utilities
      - *e*. Maintenance materials
      - *f*. Operating supplies
      - *g*. Royalties and rentals
    - 2. Labor
      - *a*. Direct operating labor
      - *b*. Operating supervision
      - *c*. Direct maintenance labor
      - *d*. Maintenance supervision
      - *e*. Payroll burden on all labor charges
        - (1) FICA tax
        - (2) Workmen's compensation coverage
        - (3) Contributions to pensions, life insurance, etc.
        - (4) Vacations, holidays, sick leave, overtime premium
        - (5) Company contribution of profit sharing
  - *B*. Indirect production costs
    - 1. Plant overhead or burden
      - *a*. Administration
      - *b*. Indirect labor
        - (1) Laboratory
        - (2) Technical service and engineering
        - (3) Shops and repair facilities
        - (4) Shipping department
      - *c*. Purchasing, receiving, and warehousing
      - *d*. Personnel and industrial relations
      - *e*. Inspection, safety, and fire protection
      - *f*. Automotive and rail switching
      - *g*. Accounting, clerical, and stenographic
      - *h*. Plant custodial and plant protective
      - *i*. Plant hospital and dispensary
      - *j*. Cafeteria and clubrooms
      - *k*. Recreational activities
      - *l*. Local contributions and memberships
      - *m*. Taxes on property and operating licenses
      - *n*. Insurance—property, liability
      - *o*. Nuisance elimination—waste disposal
    - 2. Depreciation
  - *C*. Contingencies
  - *D*. Distribution costs
    - 1. Containers and packages
    - 2. Freight
    - 3. Operation of terminals and warehouses
      - *a*. Wages and salaries—plus payroll burden
      - *b*. Operating materials and utilities
      - *c*. Rental or depreciation

* Adapted from R. H. Perry et al., "Chemical Engineers' Handbook," 4th ed., pp. 26-27, McGraw-Hill Book Company, New York, 1963.

Jelen, F. C. Cost and Optimization Engineering
McGraw-Hill Book Co. (1970)

Table 3-10 Continued
Components of Total Product Cost

- II. General expense
  - *A*. Marketing or sales costs
    - 1. Direct
      - *a*. Salesmen salaries and commissions
      - *b*. Advertising and promotional literature
      - *c*. Technical sales service
      - *d*. Samples and displays
    - 2. Indirect
      - *a*. Sales supervision
      - *b*. Travel and entertainment
      - *c*. Market research and sales analysis
      - *d*. District office expenses
  - *B*. Administrative expense
    - 1. Salaries and expenses of officers and staff
    - 2. General accounting, clerical, and auditing
    - 3. Central engineering and technical
    - 4. Legal and patent
      - *a*. Within company
      - *b*. Outside company
      - *c*. Payment and collection of royalties
    - 5. Research and development
      - *a*. Own operations
      - *b*. Sponsored, consultant, and contract work
    - 6. Contributions and dues to associations
    - 7. Public relations
    - 8. Financial
      - *a*. Debt management
      - *b*. Maintenance of working capital
      - *c*. Credit functions
    - 9. Communications and traffic management
    - 10. Central purchasing activities
    - 11. Taxes and insurance

Jelen, F. C. Cost and Optimization Engineering
McGraw-Hill Book Co. (1970)

Reprinted with permission

REFERENCES

1. Jelen, Cost and Optimization Engineering. McGraw-Hill Book Co.
2. American Association of Cost Engineers, Cost Engineers' Notebook.
3. Guthrie, K. M., Managing Capital Expenditures for Construction Products. Craftsman Book Co.
4. Faulkner, Project Management with CPM. R. S. Means Co.
5. Ostwald, P. F., Cost Estimating for Engineering and Management. Prentice-Hall.
6. Moder and Phillips, Project Management with "CPM" and "PERT". Van Nostrand -Reinhold.
7. Hackney, J. W., Control and Management of Capital Projects. John Wiley and Sons.
8. American National Standard Industrial Engineering Terminology--Cost Engineering. ANSI Z94.2 - 1972. American Institute of Industrial Engineers, and the American Society of Mechanical Engineers.
9. Building Estimator's Reference Book. The Frank R. Walker Co.
10. Vernon, Realistic Cost Estimating for Manufacturing. Society of Manufacturing Engineers.
11. Guthrie, K. M., Process Plant Estimating, Evaluation and Control. Craftsmen Book Co.
12. Clark and Lorenzoni, Applied Cost Engineering. Marcel Dekker, Inc.

4. ECONOMIC EQUIVALENCE AND PROFITABILITY

J. W. (Jim) Higginbotham
St. Regis Paper Company, League City, Texas

DEFINITION OF TERMS as used in this chapter is

Venture worth calculation differs from the Discounted Cash Flow Rate of Return in the manner of applying the interest rate. The DCFRR uses a trial and error method to determine the interest rate, whereas the Venture Worth Method employs a predetermined interest rate. For the latter a positive Venture Worth represents a premium present worth over the predetermined interest rate, and a negative result means that the predetermined interest is not realized.

Compound interest involves interest earning interest and in tables is referred to as Amount of a Dollar, or sometimes Amount of One.

Depreciation represents an expense which allocates a first cost to a yearly cost for wear and tear, and obsolescence. The three principal methods allowed by the government for the purposes are discussed fully.

Net profit is profit *after* taxes.

The information for this chapter was obtained from a booklet "Discounted Cash Flow for Investment Decision" published by Financial Publishing Company supplemented with an article by Mr. John Rodgers for Reliance Universal Company, Houston, Texas.

In this chapter all expended sums of money are indicated by a negative sign, and income is indicated by a positive sign.

EVALUATING PROJECTS requires that many angles must be explored.

- Are raw materials available in sufficient quantities?
- Are the raw materials available from more than one source?
- Has the price of the raw materials risen a great amount over a short period of time?
- Is the cash requirement for the project available from current funds or is the cash required available by a loan at a reasonable rate of interest?
- Is the product marketable at the present time or will a market need be developed?
- Are the uses for the product known or will research to establish uses be one of the requirements for sale of the product in quantities to be produced?
- Is the product now manufactured by any other corporation?
- Will plants or machinery be costly to maintain?
- Are standard machines for producing the product available?
- Are exotic materials of construction required as piping materials, tank or conveyor construction materials?
- Are patent rights available?
- Is the type of manpower available that is required for production?
- Is the manufacturing process compatible with the present processes and does the present management have enough experience to direct the operations?
- If a large amount of construction is necessary, can the contractors complete the project on schedule?

SIMPLE INTEREST is a charge applied only to the amount of the debt for a single time period.

As an example suppose $100 borrowed for 2 years at 6% per year. Then at the end of two years $112 would be due.

COMPOUND INTEREST is interest in which interest is credited to the principal each period and interest is earned on interest. The future value of a present amount A for n periods at an interest rate of i per period expressed as a decimal is

$$S = A(1 + i)^n$$

In engineering economy a year is usually taken as the period of time so that compounding is done once a year. Some analysts use continuous interest for which interest is compounded continuously, that is for each infinitesmal period of time.

SINKING FUND account can be set up to which a fixed amount is paid at the end of each period which with the accumulated interest amounts to a prescribed sum at a set time.

An advantage of a sinking fund is that cash can be provided over a longer period of time to eliminate the possible difficulty of providing the entire amount at once. One disadvantage of a sinking fund is that the rate of return obtained in the sinking fund may be less than the rate of return that might be realized by reinvestment within the company.

In terms of the terminology used in Chapter 2 let R be a uniform end of year payment to a sinking fund and let S be the amount required at the end of n years at i per year expressed as a decimal. Then by item 5 Table 1-1 in Chapter 1

$$S = R\,\frac{(1 + i)^n - 1}{i} = RF_{RS,i,n} = RF_{RP,i,n}\,F_{PS,i,n}$$

Example. Replace a fork truck costing $15,000 at the end of 5 years with a sinking fund yielding 5% compounded annually.

To find R, the end-of-year payment to the sinking fund, the above relationship gives

$$15{,}000 = R\,\frac{(1.05)^5 - 1}{.05}$$

$$R = \$2{,}715$$

That is $2,715 at the end of each year will yield $15,002 when compounded annually at 5% for 5 years.

If the firm had been netting 15% per year on investment then

$$S = R\,\frac{(1 + i)^n - 1}{i} = 2{,}715\,\frac{(1.15)^5 - 1}{.15} = \$18{,}306$$

That is $18,306 would have been realized instead of $15,002. Thus there would be a gain of 18,306 - 15,002 = $3,304 by re-investment in the firm.

Most firms can borrow cash using a delay payment period at the front of the payment schedule where only the interest is paid for a few years, and after that the payments are started on the principal.

In some instances it may be advantageous for cash flow position to delay payment on the principal until cash can be realized on the investment and sometimes enough can be realized to have a profit above the interest paid.

PROFITABILITY ANALYSIS. There are several ways to evaluate the profitability of projects and four methods can be mentioned

·Percent R.O.I. In full this stands for Percent Return on Original Investment. See Chapter 1.

·Payout time is the time required to recover the investment. See Chapter 1.

·Discounted cash flow rate of return, DCFRR, also known as interest rate of return, profitability index, investor's method, or true rate of return. See Chapter 1.

·Net present value, NPV, or "venture worth" uses a fixed rate of return but reduces value at the prescribed rate of return. See Chapter 1.

The pay out method has two weaknesses one of which is the fact that only the payout period is considered when calculating the earnings. The second is that the payout method fails to consider the pattern of earnings.

EXAMPLE

Year		Project A	B	C
0	Investment	$-100,000	$-100,000	$-100,000
1	Cash Return	50,000	50,000	10,000
2	Cash Return	50,000	50,000	90,000
3	Cash Return	50,000	50,000	50,000
4	Cash Return	50,000	50,000	30,000
5	Cash Retutn	0	50,000	0

The payout time for each of the above is 2 years and this would make the projects equally attractive. However, when we analyze the above we find that project "A" fails to show a profit after year 4. Project "B" continues to show a profit after year 4.

Project "C" recovers only $10,000 the first year so $40,000 of profit or cash flow is delayed until the second year and this means $40,000 cannot be utilized in project "C".

Another disadvantage of projects "A" & "C" is the amount of total earning. Project "A" earns $200,000 total, project "C" earns $180,000 total, and project "B" earns $250,000 for the same period and continues to earn.

A DCFRR calculation for the three projects yield

Project A	35%
Project B	41%
Project C	26%

Project B is the most attractive by that criterion.

Many factors enter into and determine the rate of return for projects.

- Risk of sale of the product and the price
- Increase in cost of raw material and its availability
- Competition against superior or lower prices products
- Maintenance expenses if excessive

SALES VOLUME and selling price are particularly important in profitability studies since they can make or break a project. The ratio of profit to sales is an important figure in finance. Thus if sales are 184.3 Mega$ and profit is 18.28 Mega$, the ratio is

$$\frac{18.28}{184.3} = 9.91\%$$

That is it takes almost $10 of sales to generate $1 of profit.

DEPRECIATION METHODS were discussed in Chapter 1. The methods covered were

- Straight line, SL
- Declining balance, DB
- Sum-of-the-years digits, SD
- Double declining balance, DDB

The methods give different depreciation schedules as the following example will show.

Suppose for an example that a machine costs 100,000 when new and lasts 16 years with no salvage value.

For straight line depreciation the depreciation will be constant each year at

$$\frac{100,000}{16} = 6,250 \ \$/yr$$

For sum-of-the-years-digits method of depreciation, noting that the sum of the digits from 1,2,3 ··· + 16 is 136, there results

Year	Depreciation	%
1	100,000 x (16/136) = 11,800	11.8
2	100,000 x (15/136) = 11,000	11.0
3	100,000 x (14/136) = 10,300	10.3
4	100,000 x (13/136) = 9,600	9.6
...	...	...
14	100,000 x (3/136) = 2,200	2.2
15	100,000 x (2/136) = 1,500	1.5
16	100,000 x (1/136) = 700	0.7
	cumulative =100,000	

For the double declining balance the factor is here given by

$$\frac{2}{16} = 0.125$$

and there results

Year	Book Value	Depreciation	%
1	100,000	100,000 x 0.125 = 12,500	12.5
2	87,500	87,500 x 0.125 = 10,938	10.9
3	76,562	76,562 x 0.125 = 9,570	9.6
4	66,992	66,992 x 0.125 = 8,374	8.4
...	...	...	...
14	15,422	15,422 x 0.125 = 2,203	2.2
15	13,494	13,494 x 0.125 = 1,928	1.9
16	11,807	11,807 x 0.125 = 1,687	1.7
		cumulative = 88,193	

The double declining balance method gave an accumulated depreciation of \$88,193 leaving \$11,807 not depreciated. The circumstance can be surmounted by shifting to straight line depreciation at any chosen point. Thus suppose the shift to straight line is made at the end of year 9 when the book value, the underpreciated value is 34,362. Then the straight line depreciation with 8 years remaining is 34,362 ÷ 8 = \$4,295 and the tabulation becomes

Year	Book Value	Double Declining Balance Depreciation	Straight Line Depreciation	Cumulative Depreciation
1	100,000	12,500		12,500
2	87,500	10,938		23,538
3	76,562	9,570		33,008
...	...	...		...
8	39,270	4,908		65,638
9	34,362		4,295	69,933
10	30,067		4,295	74,228
...	...		...	...

Year	Book Value	Double Declining Balance Depreciation	Straight Line Depreciation	Cumulative Depreciation
14	12,887		4,295	91,408
15	8,592		4,295	95,703
16	4,297		4,297	100,000

Thus the switch to straight-line method of depreciation permits the entire investment of $100,000 to be written off.

The previous compilations have neglected salvage value. When salvage value exists and is decreed by government rules, then it is never permissible to depreciate more than the depreciable value, which is first cost less salvage value. Otherwise expressed the book value must not fall below the salvage value. If the project is sold for its salvage value at the end of the project, there is no gain or loss at that time and no tax consideration. If sold for more than the salvage value there is a profit subject to tax. Conversely if sold for less than the salvage value, there is a loss which reduces the tax.

SOME EXAMPLES. Some practical examples of economic comparisons and profitability follow.

ADD TO EXISTING FACILITY. Existing facility number 1 generates a cash flow of $80,000 per year and an addition to this facility at a cost of $100,000 is contemplated. The existing facility realizes 15% on investment, and the addition will generate an increase of $21,000 cash flow for a period of 16 years. The Net Present Value, or Present Worth of the addition and the extra income is by Chapter 1, Table 1-1.

$$NPV = -100,000 + 21,000F_{RP,i,16}$$

$$NPV = -100,000 + 21,000 \frac{(1 + i)^{16} - 1}{i}$$

For the Net Present Value to be zero, i becomes the discounted cash flow rate of return and

$$0 = -100,000 + 21,000F_{RP,i,16}$$

$$F_{RP,i,16} = 4.76 = \frac{(1 + i)^{16} - 1}{i}$$

The value of i can be solved by trial and error and is about 20%. Also, one can scan Tables for $F_{RP,i,16}$ and find that at 20% $F_{RP,20\%,16} = 4.73$ which is close enough. Thus the addition earns 20% rate of return or 5% more than that for the existing facility.

If the income per year is irregular, as opposed to uniform, then the present worth for each year must be calculated individually with a trial value for i until the present worth is zero. The better pocket programmable calculators have programs for finding the DCFRR so the calculation hardly needs to be done by hand any longer.

EXAMPLE OF MAINTENANCE DECISION. DCF can also be applied in a similar manner to evaluate maintenance decisions. Consider replacing a pump when two different types can be purchased and either type will perform the task.

Pump number 1 costs $13,000 with a life of 18 years with negligible maintenance.

Pump number 2 costs $10,000 with a life of 18 years, and a maintenance of $7,000 (70% of original cost) at the end of 6 and 12 years.

Let the tax rate be 46%. The pumps are used for replacement, and assume additionally that the first cost can be written off immediately for tax purposes. Likewise the maintenance cost of $7,000 per year can be written off immediately for tax purposes.

The present values of pump number 1 over pump number 2 can be shown in a tabulation. Note that the after tax *extra* cost for pump number 1 is

$$13{,}000(1 - .46) - 10{,}000(1-.46) = \$1{,}620$$

The maintenance cost after taxes is

$$3{,}500(1 - .46) = \$1{,}890$$

The discount factor is $(1 + i)^{-n}$ and can be obtained from Tables or calculated as needed. The tabulation is

Year	Extra cash flow	Discount factor 10%	Discount factor 12%	Total discounted cash flow 10%	Total discounted cash flow 12%
0	-1620	1.0000	1.0000	-1620	-1620
6	+1890	0.5645	0.5066	+1067	+957
12	+1890	0.3186	0.2567	+602	+485
				+49	-178

By interpolation the discounted cash flow is zero at about 10.4%. That is, pump number 1 returns 10.4% compared to pump number 2.

LEASE AGAINST PURCHASE. The question of when to lease and when to purchase equipment becomes very complex at times and consideration of all facets must be made.

For *purchasing* one consideration is the effect of the added capital assets on the tax structure of the Corporation and the additional bookkeeping necessary to calculate depreciation, etc., when the assets are acquired.

Records of maintenance and efficiency of the different types of equipment must be maintained so the time of replacement can be determined and provided for in the budget.

Storage and repair facilities must be provided for the equipment and the repair parts. The availibility of trained maintenance personnel must be evaluated and the time they will be required must be determined.

One of the most important considerations must be to determine how frequently the equipment will be employed in the normal working year. Another important phase of this is to evaluate the mobility of the equipment and the ability to utilize the equipment in other locations of the mill or plant sites and how charges for equipment are to be distributed.

Security maintained in the areas in which the equipment or repair parts and tools are stored can be another important phase to consider.

However, the equipment is available when required on the site and a move-in, move-out charge is eliminated.

For *leasing* some considerations are no capital assets are added to have an effect on the tax structure of the Corporation and the bookkeeping is not increased to calculate depreciation, etc., and maintain records for control.

No formal records are needed for normal maintenance requirements and efficiency determination

Storage and repair facilities need not be provided and trained maintenance personnel are not needed.

If a type of equipment is not employed frequently and for long periods of time the equipment rental will cease when the equipment is released from the project. Another important phase of this is the fact that the charge for the equipment can be transferred by the Contractor to another project with ease. If the equipment is owned, internal bookkeeping and agreement of the amount of charges are required.

Security of the equipment must be maintained only when the equipment is utilized as the equipment is on site only when required.

Availibility of equipment and the distance of haul must be considered and a contractor must be selected who can furnish the equipment to be required with consideration of the cost of move-in, move-out charges.

EXAMPLE OF LEASE VS PURCHASE. A new welding machine (gasoline/electric) has been leased for $300 per month and a new machine can be purchased for $2,500. We must now decide to purchase a new machine or continue to lease a machine. The data are

Cost of new machine, $	2,500
Cost of leasing, $/yr	3,600
Expected life of machine, yr	8
Gas, oil, maintenance, etc is cancelled as they must be provided for leased machine	0
Straight line depreciation, years total	8
Salvage value	0
Tax rate	46%

The leasing cost can be written off at once for tax purposes and after taxes is

$$3600(1 - .46) = \$1,944$$

The new machine generates a depreciation expense of 2500/8 = $312.50 which after taxes becomes

$$312.50(1 - .46) = 168.75$$

The extra cash flow by purchasing is

$$-2,500 \text{ at zero time}$$

and there is a savings over leasing at the end of each of the 8 years amounting to

$$1944 + 168.75 = \$2112.75$$

A year by year tabulation is not necessary in this example but is used as a guide to a general procedure. The discounted cash flow for the extra cash flow generated by purchase of pump number 1 is

Year	Extra cash flow	Discount factor 90%	Discount factor 80%	Total discounted cash flow 90%	Total discounted cash flow 80%
0	-2,500	1.0000	1.0000	-2,500	-2,500
1	+2,112.75	0.5263	0.5556	1,112	1,174
2	+2,112.75	0.2770	0.3086	585	652
3	+2,112.75	0.1458	0.1715	308	362
4	+2,112.75	0.0767	0.0953	162	201
5	+2,112.75	0.0404	0.0529	85	112
6	+2,112.75	0.0213	0.0294	45	62
7	+2,112.75	0.0112	0.0163	24	34
8	+2,112.75	0.0059	0.0091	12	19
				-167	+116

By interpolation the discounted cash flow is zero at about 84.1% in favor of purchasing a new pump.

EXAMPLE OF POPULATION GROWTH. A person starts a rabbit farm with 1,000 rabbits. Rabbits increase by 10% per year. Each year 50 rabbits must be sold to pay expenses. How long will it take to reach a population of 4,000 rabbits?

Set 500 rabbits aside to generate the 50 rabbits per year sold for expenses. The other 500 rabbits must increase 7-fold to 3500 at 10% per year for n years. From Tables or by trial and error

$$(1.10)^n = 7$$

and n = 20.4 years.

EXAMPLE OF IMMIGRATION. Assume 300,000 new people enter a country each year and after entering increase by 1% per year. If none die, how many people will be added from immigration in a period of 50 years?

4-16

Using the methods of Chapter 1, Table 1-1

$$S = RF_{RS,1\%,50} = R\frac{(1 + i)^n - 1}{i} = 300,000\frac{(1.01)^{50} - 1}{.01}$$

S = 19,338,900 people at end of 50 years

EXAMPLE OF INVENTORY TURNOVER. A store has a net profit of 5% on each sale and an average turnover of 3 times a year. If the turnover increases to 5 times a year, how will the profit be increased?

At 3 turnovers per year

$(1.05)^3 = 1.1576$ or 15.76% profit

At 5 turnovers per year

$(1.05)^5 = 1.2763$ or 27.63% profit

The profit would increase from 15.76% to 27.63%

EXAMPLE OF RATE OF GROWTH. If a corporation estimates a market for 5,000,000 units per year of a product with a present manufacturing capacity of the plant at 150,000 units per year, and the corporation wants to expand to 5,000,000 units per year within the 17 years (the life of the patent) what will be the rate of expansion per year?

If 150,000 must grow to 5,000,000 then 1 must grow to 5,000,000/150,000 = 33.33. We can scan Tables and find that at 22% 1 grows to 29.38 in 17 years, and at 23% 1 grows to 33.75 in 17 years. Thus the rate of growth is about 23% per year.

EXAMPLE ON POLLUTION. The present level of an impurity is 100 and the allowable level is 300. Each piece of added equipment will increase the amount of pollution 2%. How many pieces of additional equipment can be added?

$$(1.02)^n = \frac{300}{100}; \text{ or } n \log 1.2 = \log 3$$

for which n = 55.5. Thus 55 new pieces of equipment can be added without exceeding the allowable pollution level.

ECONOMICS FOR CAPITAL PROJECTS OR REPLACEMENTS. Economic justification for construction or replacement of a project should be viewed, in addition to other respects, from the aspect of the cost of the project in terms of the impact the cost imparts to the method of obtaining revenue. Most corporations obtain revenue from the sales of products, so the impact the cost will have on sales revenue must be considered.

When calculating Discounted Cash Flow the cost of money, profit, depreciation, etc. must be established to a high degree of accuracy and these figures can be applied in other methods of calculating risk and profitibility of projects.

One of the first steps in determining the profitibility in the impact on sales method is to examine the present percentage of sales to profit. This can be done by dividing the net profit for the year by the sales volume in dollars for the year. If we assume the sales volume to be \$ 184,300,000 and the net profit to be \$ 18,280,000 we find the percentage to be:

$$\frac{\$\ 18,280,000}{\$\ 184,300,000} = 9.91\%$$

From the calculation above the ratio of net profit to sales is found to be 9.91% and by applying this percentage, an amount of sales required to off set the expenditure can be found.

For each one dollar (\$-1) expended the amount of sales required is found to be:

$$\$\ \frac{1}{.0991} = \$\ 10.09$$

The calculation above illustrates the importance of the corporation saving comparatively small amounts of money as well as larger amounts. When viewed from the evaluation aspect and applied to the project several determinations must be made some of these are:

1. The amount of sales volume in dollars per year that will be generated.
2. The amount of net profit realized from the additional sales volume.
3. The amount of operating expense to be eliminated by more efficient operations.
4. The duration in time of the increase in production or duration in time of the savings in expense.
5. The amount of sales volume in dollars per year required to offset the expenditure.

As an example:

Estimated ratio net profit to sales = 10.%

Estimated cost of project = $ 45,000,000

Estimated sales volume required = $ 450,000,000

$$\frac{\$\ 45,000,000}{10} = \$\ 450,000,000 \text{ total}$$

Estimated sales volume generated per year = $ 37,000,000

Estimated net profit generated per year = $ 3,700,000

Depreciation (16 years straight line) = $ 2,812,500 per year

$$\frac{\$\ 45,000,000}{16} = \$\ 2,812,500$$

Break even point (net profit sales only) = 12.16 years

$$\frac{\$\ 45,000,000}{3,700,000} = 12.16 \text{ years}$$

Break even point (net profit + depreciation) =

$$\frac{45,000,000}{3,700,000 + 2,812,500} = 6.9 \text{ years}$$

When the "Discounted Cash Flow" method is applied as "Venture Worth" using the same 10% factor to determine if the project is feasible we find:

YEARS	CASH FLOW	DISCOUNT FACTOR INTEREST RATE 10%	TOTAL DISCOUNTED CASH FLOW 10%
0	(-$ 45,000,000)	1.0000	($-45,000,000)
	Net profit + Depreciation $3,700,000 + $2,812,500 = $6,512,500		
1.	$6,512,500	.9091	$ 5,920,514
2.	6,512,500	.8264	5,381,930
3.	6,512,500	.7513	4,892,841
4.	6,512,500	.6830	4,448,038
5.	6,512,500	.6209	4,043,611
6.	6,512,500	.5645	3,676,306
7.	6,512,500	.5132	3,342,215
8.	6,512,500	.4665	3,038,081
9.	6,512,500	.4241	2,761,951
10.	6,512,500	.3855	2,510,569
11.	6,512,500	.3505	2,282,631
12.	6,512,500	.3186	2,074,883
13.	6.512,500.	.2897	1,886,671
			$ 46,260,241
14.	6,512,500	.2633	1,714,741
15.	6,512,500	.2394	1,559,093
16.	6,512,500	.2176	1;417,120
			$ 50,951,195

As shown above the project would earn the required 10% profit and the capital investment recovered the thirteenth (13) year with $ 1,260,241 or 2.8% of the initial $ 45,000,000 investment as a surplus.

If the calculation is completed until the sixteenth (16) year when the project has been totally depreciated the 10% profit would be earned and the initial investment recaptured with a \$ 5,951,195 surplus.

The above is an example of a good investment, however, in another example if money was expended and contributed nothing to the profit or did not reduce operating expenses we have a different example.

Money expended = \$ 150,000

Sales needed to break even = \$ 1,500,000

Estimated ratio net profit to sales = 10%

When the example above is viewed in the effect on sales, profit and the loss of revenue which could have been realized if the profit had been invested in a project yielding 10% profit and even greater impact on earnings is noted.

$$\$\ 150,000 + (.10 \times 150,000) = \$\ 150,000 + \$\ 15,000 = \$\ 165,000$$

$$\frac{\$\ 165,000}{.10} = \$\ 1,650,000 \text{ additional sales required to break even}$$

When viewed by using the above calculation the sales required to off set an additional expenditure during the year of \$ 150,000 + \$ 15,000 we find as above \$ 1,650,000. When this is applied to a corporation earning 10% on sales of \$ 182,800,000 or \$ 18,280,000 the impact of the expenditure becomes:

$$\frac{18,280,000}{182,800,000} = 10\% \text{ net profit}$$

$$\frac{165,000}{18,280,000} = 0.009 \text{ or approximately } 1\% \text{ of the total net profit of the corporation.}$$

The ratio of the smaller gain in sales can present another picture as can be seen above, \$ 1,650,000 of the sales revenue must be expended for the \$ 150,000 spent. This represents

$$\$ \frac{1,650,000}{182,800,000} = 0.009 \text{ or approximately 1\% of total sales.}$$

If a corporation with the above sales volume of \$ 182,800,000 has a healthy net gain in sales of 12% for the year or \$ 21,936,000 an expenditure of only \$ 150,000 for a non productive project would result in a reduction of $\$ \frac{1,650,000}{21,936,000} =$ 0.075 or 7.5% of the net gain in sales. This would for all practical purposes reduce the net gain in sales to 11% or \$ 20,108,000 from the 12% or \$ 21,936,000 gain, as 7.5% of the additional sales would be off set by the nonproductive expenditure.

The examples above illustrate the importance of the evaluation of all projects to be under taken by corporations. The examples also illustrate the importance of efficient operations and budget planning when the impact on sales ratio of profits and the impact on net profits form a comparatively small unnecessary expenditure is noted.

When considering the growth of a corporation a 12% increase in sales appears to be a healthy rate of growth. However, when a \$ 150,000 of unnecessary expenditures can lower 1% of the sales growth and \$ 2,193,600 can negate the entire growth in sales, the careful evaluation of expenditures can be seen as of greater importance.

$$\$ \frac{2,193,600}{.10} = \$ 21,936,000 \text{ Sales required to realize \$ 2,193,600 net profit at 10\% net return.}$$

The importance of implementing more efficient conditions or methods where possible can readily be seen when the fact is noted that a saving of \$ 150,000 in operating expense is the equivalent of a 1% growth in sales for a corporation with a sales volume of \$ 182,800,000. The above becomes even more significant if the operating expenses are average or approximately 85% of sales or revenue.

$$.85\% \times \$\ 182,800,000 = \$\ 155,380,000$$

When the realization that with a comparatively small saving, $\frac{150,000}{155,380,000} = 0.009$ or approximately one tenth of one percent, of operating expense the equivalent of a 1% growth in sales is realized the saving becomes of even more importance.

SCHEDULING A LOAN PAYMENT. A loan of \$1,000,000 is borrowed with interest at 8% per year or 2% per quarter years. The loan will be repaid over a scheduling period of 25 years. During the first 5 years, 20 periods, only interest will be paid. During the next 20 years, 80 periods, the loan balance will be reduced uniformly and interest paid on the loan balance.

Here 1,000,000 is repaid uniformly over the last 80 periods at 1,000,000/80 = \$12,500. The payment schedule is

Quarter	Interest per period	Principal per period	Payment per period	Loan Balance
1-20	20,000	0	20,000	1,000,000
21	20,000	12,500	32,500	987,500
22	19,750	12,500	32,250	975,000
...	...	...	...	...
50	12,750	12,500	25,250	625,000
51	12,500	12,500	25,000	612,500
...	...	...	...	...
99	500	12,500	13,000	12,500
100	250	12,500	12,750	-0-
	$1,210,000	$1,000,000	$1,810,000	

At times it might be advantageous to delay payments toward principal to maintain a better cash flow position and the amount paid in interest for the delay period must be weighted in this decision. In this example the interest for the 5 years is 49.38% of the interest of that of the remaining 20 years of the loan. The interest for the first 5 years also is 40% of the loan so if the loan was $50,000,000 the interest for the 5 year delay period would be $20,000,000 or quarterly payments of $1,000,000 or $4,000,000 per year.

The above $20,000,000 paid over the 5 year period would then be an addition to the $50,000,000 cost of the project and for Discounted Cash Flow Calculations the $20,000,000 would be paid by the profit from the project for the first 5 years. Any cash expended while the project was under construction would be included with the loan.

EXAMPLE OF LOAN SCHEDULE. For an example take the installation of a paper machine and stock preparation area. The following apply

Total installed cost	$55,000,000
Amount of loan	$50,000,000
Cash outlay during construction	$ 5,000,000
Repayment of loan	
First 5 years interest only.	
Thereafter paid off in 80 quarterly payments at beginning of 6th year	
Interest rate	8%
Depreciation method	straight line
Depreciation years	16
Depreciation, $/yr 55,000,000/16	$ 3,437,500
Net profit after interest and loan payments	
Year 1, $/yr	$ 1,500,000
Years 2-25, $/yr	$ 3,300,000
Cash flow = depreciation + net profit	
Year 1, 3,437,500 + 1,500,000	$ 4,937,500
Years 2-16, 3,437,500 + 3,300,000	$ 6,737,500
Years 17-25, 0 + 3,300,000	$ 3,300,000

The first part of the cash flow calculation is a restricted view to accentuate paying only interest on the loan for the first 5 years. The project is judged from a commitment of \$20,000,000 in interest (5 years at \$4,000,000/yr). Although it is true the interest will not be due at zero time but will be spread out in the future, the view here is to consider the \$20,000,000 as a present commitment of the project. For the first 5 years the calculation is

Year	Reference	Cash flow	Discounted at 17%	Discounted at 18%
0	-20,000,000		-20,000,000	-20,000,000
1		4,937,500	4,220,081	4,184,531
2		6,737,500	4,921,744	4,838,873
3		6,737,500	4,206,895	4,100,443
4		6,737,500	3,595,804	3,475,203
5		6,737,500	3,072,974	2,944,961
			+ 17,498	-455,989

The present worth will be zero at about 17.04%. This says in effect that if the total interest payments for the first 5 years, \$20,000,000 were swapped for the first 5 years of the project, the discounted cash flow rate of return would be about 17.04%

The second part of the cash flow calculation goes from years 6 through 25 and the reference point here is a 55,000,000 investment. All discounting is done with respect to the beginning of year 6. The calculation is

Year	Reference	Cash flow	Discounted at 8%	Discounted at 10%
	-55,000,000		-55,000,000	-55,000,000
6		6,737,500	6,238,251	6,125,061
7		6,737,500	5,776,059	5,567,870
...		...	...	...

15	6,737,500	3,120,810	2,597,306
16	6,737,500	2,889,714	2,361,494
17	3,300,000	1,310,430	1,051,380
18	3,300,000	1,213,410	956,010
...	...	...	...
24	3,300,000	764,610	539,550
25	3,300,000	707,850	490,380
		+1,939,281	-4,579,217

The present worth will be zero at about 8.60%. This says that if the project is regarded as an investment of 55 Mega$ at the beginning of the 6th year, it will have a discounted cash flow rate of return of 8.60% from the period starting at the beginning of the 6th year and ending at the end of the 25th year.

One of the major advantages in the process of delaying the payments toward the loan principal is to take advantage of the cash flow realized from depreciating the capital assets during the time. There is a possibility, however, of the delay of payments being a detriment if the debt position of the Corporation is good (very few debts outstanding) and a major expansion or major maintenance program is contemplated within a 4 or 6 year period. A possibility exists under these circumstances that the credit rating of the Corporation would be affected. The Corporation may then be in a better position to repay the debt at a more accelerated rate to insure a more advantageous position in credit rating and solvency. This may insure the ability to borrow sufficient funds at a later date to allow expansions or maintenance at a critical time in the future.

When contemplating the life of a Corporation, plans must be projected to allow the Corporation to continue indefinitely and not be affected by changes in personnel or any other short term occurrence. Cycles in the change of personnel, business climate (cost of money expenses, etc.) should be projected and planned so that the Corporation will have a healthy growth forever.

PROFITABILITY OF A THEOTETICAL PAPER MACHINE. A paper machine has a production of 375 ton/day for 350 day/yr at sale price of $281.50 per ton. Gross income per year is

$$375 \times 350 \times 281.50 = \$36,946,875 \quad \$/yr$$

After expenses, but before taxes, only 15.16% remains or

$$.1516 \times 36,946,875 = \$5,601,146 \quad \$/yr$$

After allowing $2,238,980 for taxes, the net income after taxes is

$$5,601,146 - 2,238,980 = \$3,362,166 \quad \$/yr$$

The capital cost for the machine is $34,000,000 and it will be depreciated at Straight-line depreciation over 16 years or at

$$34,000,000/16 = \$2,125,000 \quad \$/yr$$

The cash flow is the sum of net profit after taxes plus depreciation and is

$$3,362,166 + 2,125,000 = \$5,487,166 \quad \$/yr$$

The return on original investment for the machine is

$$\frac{5,487,166}{34,000,000} = 16.1\%$$

The payback time will be

$$\frac{34,000,000}{5,487,166} = 6.20 \text{ years}$$

All of the above calculations have neglected the 10% credit for 1976 which amounts to

$$34,000,000 \times .10 = \$3,400,000$$

When the 10% tax credit is considered, the return on original investment becomes

$$\frac{5{,}487{,}166}{34{,}000{,}000 - 3{,}400{,}000} = 17.9\%$$

The payback time when considering the tax credit is

$$\frac{34{,}000{,}000 - 3{,}400{,}000}{5{,}487{,}166} = 5.58 \text{ years}$$

Cash flow for the first year is

Profit after tax	$3,362,166
Depreciation	2,125,000
10% Investment tax credit	3,400,000
	$8,887,166

and for the other years is

Profit after tax	$3,362,166
Depreciation	2,125,000
	$5,487,166

The venture worth of the project at 10% rate of return is + $12,036,000 from the following calculation

Year	Amount	10% present worth factor	Present Worth
0	-34,000,000	1.0000	-34,000,000
1	8,887,166	0.9090	8,078,434
2	5,487,166	0.8264	4,534,594
3	5,487,166	0.7513	4,122,508
...	...	...	...
15	5,487,166	0.2394	1,313,628
16	5,487,166	0.2176	1,194,007
			+12,035,904

5. COST REPORTING AND CONTROL

D. L. Anderson
Brown and Root, Inc., Houston, Texas

OUTLINE for this chapter follows:

I. Project budget
   A. Introduction
      1. Project involving design - construct
      2. Organization
         a. Proper set up
         b. Proper control
      3. Development of budget
         a. Responsibilities
         b. Influences
   B. Organizational
      1. Negotiation
         a. Selection of Personnel
         b. Staff functions

2. Coordination of effort
3. Engineering
4. Construction

C. Estimate preparation
   1. Responsibilities
   2. Types of estimates
   3. Timeliness of budget completion

D. Variance
   1. Individual
   2. Cumulative

E. Cash flow
   1. From schedule or historical
   2. Comparison

F. Budget
   1. On computer
   2. Parallel's estimate

G. Cost integrated schedule
   1. Manhour cost and time compared
   2. Physical percent complete used
   3. Integrated at summary level

II. The chart of accounts

A. Reason
   1. Well defined controllable breakdown
   2. Measurement

B. Development
   1. During estimating stage by cost engineer--cooperations
      a. Estimating
      b. Scheduling
      c. Engineering
      d. Data processing
   2. Client interests
      a. Special needs of client
      b. Information
      c. Flexible

3. Battery limits-areas
    a. Physical boundaries
    b. Process boundaries
    c. Not by system

C. Coding structure
1. Primary codes
    a. Cost indicator
    b. Area
    c. Activity
    d. Cost class-accounting
2. Supplementary codes
    a. Equipment
    b. Client special requirements

III. Reports

A. Kinds of
1. Weekly labor report
2. Payroll labor distribution
3. Crew size distribution
4. Job classification report
5. Summary weekly labor report
6. Summary quantity and manhour report
7. Level 1 exception report
8. Management summary
9. Weighted progress report
10. Detail cost ledger
11. Open commitment worksheet
12. Monthly cost comparison
13. Management summary cost comparison

IV. Engineering and construction cost analysis

A. Physical progress
1. Establish standard drawing milestone schedule
2. Assign weight (manhours) to each drawing

3. Level of completion allowed at certain stages
    a. Design only--20% complete
    b. Drawn--60% complete
    c. Office clerk--75% complete
    d. Client approval--85% complete
    e. Issue--95% complete
4. Physical progress % = $\frac{\text{Earned weight}}{\text{Total drawing weight}}$
5. Productivity = $\frac{\text{Weighted physical \% complete}}{\text{\% of current budgeted manhours}}$

B. Graphic display
1. Progress control chart
2. Productivity profile curve

V. Procurement costs

A. Operations
1. Purchasing department--cost conscious
    a. Buying
    b. Expediting
    c. QA/QC
    d. Traffic

B. Engineered materials
1. Vendor evaluation
2. Specifications
3. Inquiry
4. Inquiry to vendor
5. Recieve quotations
6. Commercial evaluation
7. Technical evaluation
8. Client approval
9. Requisition
10. Prepare purchase order of substitute
11. Expediting
12. Traffic

13. Receipt
14. Warehousing
15. Installation

C. Bulk materials
1. Vendor evaluation
2. Specifications
3. Take-offs
4. Inquiry
5. Quotations
6. Commercial evaluation
7. Prepare purchase order
8. Expediting
9. Traffic
10. Receipt
11. Warehousing
12. Installation
13. Disposal of surplus

This chapter is prepared in connection with experience at Brown and Root, Inc. There are seven basic operating divisions within the corporate structure and the cost control procedures can vary within the operating divisions. A common basic system, however, can be used to fit various types of service or needs of a client.

At Brown and Root, Inc. the responsibility for cost control lies with each individual project. Cost report generation, maintenance, analysis, forecasting, and budget updating is the responsibility of the field cost engineer aided by the home office division cost manager. The responsibility remains with the project, however, and the cost report projections or forecasts and analysis are done at the project site by project personnel. The project is responsible for the staffing of cost engineers and selection of cost reports.

The projects are restricted to the system through corporate procedure manuals which dictate a standard based upon good accounting practice. For example, all input for both accounting and cost are from a single source document. The foreman's daily time sheets are the basic source documents for the payroll and payroll distribution. The same applies to the voucher payables for materials and sub-contracts to assure a reconciliation of cost reports with accounting statements. It is also a labor saver by requiring only a single audit function.

The cost control system provides five basic functions

- 1. Provides a control system utilizing a budget
- 2. Provides control of cost during design and construction
- 3. Provides factors or necessary means to determine productivity
- 4. Provides feedback to the client and management to produce effective and desired control
- 5. Provides control procedures to reduce excess material

The elements of cost control are

- 1. What is cost control?
- 2. What are the essential duties of the cost engineer?
- 3. How does management utilize cost engineering?

A simple definition states that cost control is the application of procedure to minimize cost in relation to a budget. Cost control begins with basic design concept and continues through completion of construction. Cost control aids engineering in evaluation of design in terms of cost to provide the best plant design for low initial cost, economic operation, and inexpensive maintenance. It is an aid to construction by providing efficient scheduling, expediting, and application of men and materials for the project combined with constant analysis of expenditures to stay abreast of cost trends in order to provide the completed project at the lowest cost in the shortest time.

ORGANIZATIONAL. A simple organizational chart is shown in Table 5-1. The viewpoint here is from that of a large construction company like Brown and Root Inc. and here referred to simply as the company. During negotiation with the client prior to actual award of a contract the project general manager, engineering project manager, and perhaps others will have already been selected. The "project general management staff members" will be headed by a "project control manager" reporting directly to the "project general manager". These discussions with the client will cover such topics as costs, budgets, estimates, schedules, material control, administration, and so on.

Upon award of the contract the cost engineer with the help of other staff members will begin setting up the cost control procedures, including the estimating and scheduling procedures, which will become part of the overall "project procedure manual". The client will be furnished the chart of accounts as it is being developed, and it will be the basic structure upon which the budget will be built.

Table 5-2 is a chart of the engineering organization. Engineering will probably have 6 to 8 months lead time over construction operations in the field. Each engineering discipline should be responsible for its portion of the estimate or budget. The cost engineer cooperates with the estimators and schedulers of each discipline to get up consistent accounts. Preprinted estimating sheets should be set up to guide each engineering discipline in preparing its budget. The next four exhibits, Tables 5-3 through 5-6 are examples of four such sheets.

Table 5-1
Simple organization chart

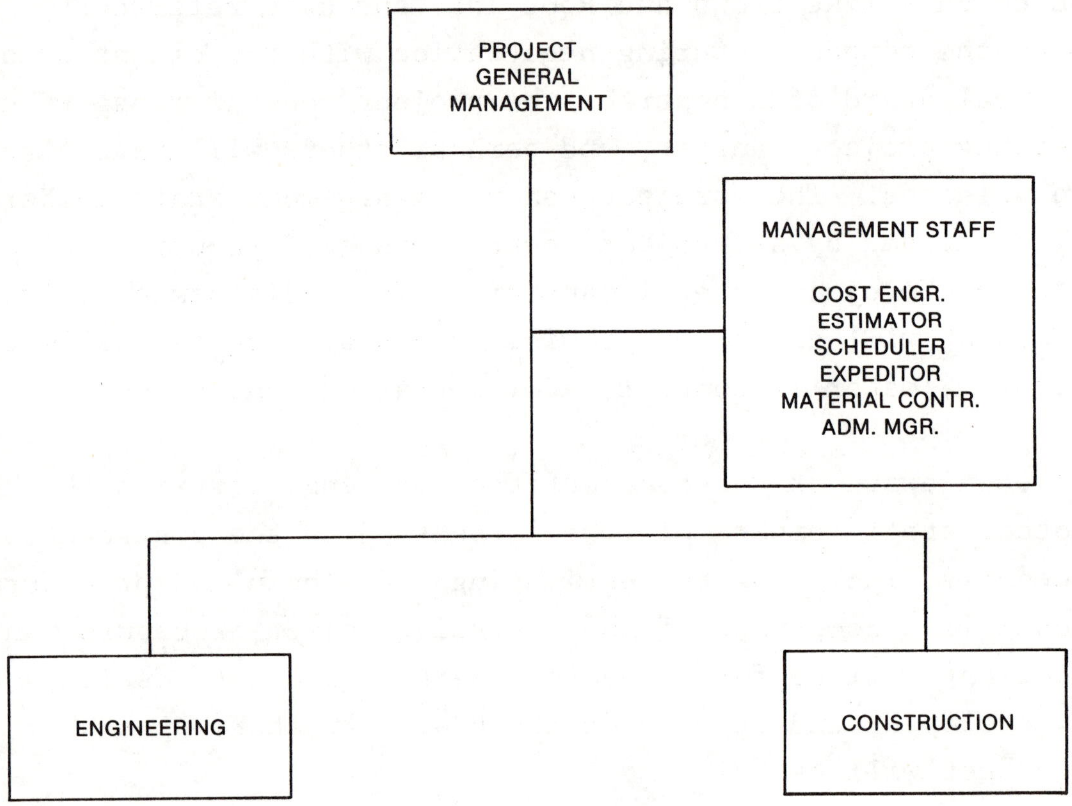

Table 5-2
Engineering organization

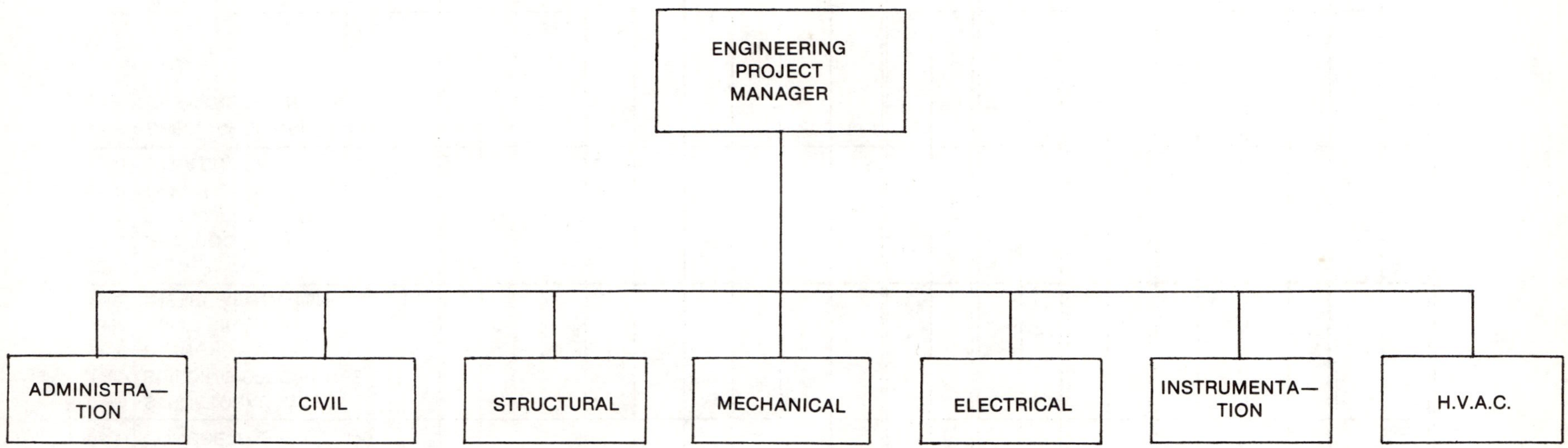

Table 5-3

SUMMARY SHEET

PROJECT ______________________ DATE ______________

BID ITEM ______________________ ESTIMATOR ______________

SUB ITEM ______________________ SHEET NO. ________ OF ________

MH	ITEM	UNIT	QUANTITY	UNIT COST						TOTAL COST					
				L	PM	M&S	EO	P&E	T	LABOR	PM/SC	M&S	EO	P&E	TOTAL
0100	SITE PREPARATION														
0200	SITE IMPROVEMENTS														
0300	AREA PREPARATION														
0700	PILING														
0800	CONCRETE														
0900	STRUCTURAL STEEL														
1000	BUILDING SPECIALTIES														
1700	PIPING														
1800	ELECTRICAL														
1900	INSTRUMENTATION														
2000	INSULATION														
2100	PAINTING														
4000	EQUIPMENT														
9100	CLEAN-UP														
9200	START-UP														
9700	SPEC. MAT'L HANDLING														
9800	CONSTRUCTION EQUIPMENT														

Table 5-4

4000 MAJOR EQUIPMENT

PROJECT ______________________ DATE ______________

BID ITEM ______________________ ESTIMATOR ______________

SUB ITEM ______________________ SHEET NO. ________ OF ________

MH	ITEM	UNIT	QUANTITY	UNIT COST						TOTAL COST					
				L	PM	M&S	EO	P&E	T	LABOR	PM/SC	M&S	EO	P&E	TOTAL
	TOWER OR PRESS. VESSEL														
	WT.														
	O.X. DIM DIA X														
	+ 12														
4010	PROCUREMENT & EQUIPMENT	EA													
	COST														
	FREIGHT	EA													
	TAXES	EA													
	+ 12														
4010	SUB TOTAL														
	+ 12														
4020	UNLOAD & HAUL	EA													
	+ 12														
4030	ERECTION														
	+ 12														
4031	SPECIAL RIGGING	EA													
	+ 12														
4032	INTERNALS	EA													
	+ 12														
4030	SUB TOTAL														
	+ 12														
4040	STRESS RELIEVE	EA													
	+ 12														
4050	X-RAY	EA													
	+ 12														
4060	TESTING	EA													
	+ 12														
4000	SUB TOTAL														

Table 5-5

0700 PILING

PROJECT ______ DATE ______

BID ITEM ______ ESTIMATOR ______

SUB ITEM ______ SHEET NO. ______ OF ______

MH	ITEM	UNIT	QUANTITY	UNIT COST						TOTAL COST					
				L	PM	M&S	EO	P&E	T	LABOR	PM/SC	M&S	EO	P&E	TOTAL
0710	SHEET PILING	LF													
0711	PERMANENT	LF													
0712	TEMPORARY	LF													
0710	SUB TOTAL	LF													
0720	UNTREAT WOOD PILE	LF													
0721	DRIVE	LF													
0722	LOAD TEST	LF													
0723	CUT-OFF	LF													
0724	STEEL CAP	LF													
0720	SUB TOTAL	LF													

Table 5-6

SUMMARY SHEET

PROJECT ______________________________ DATE ______________

BID ITEM ______________________________ ESTIMATOR ______________

SUB ITEM ______________________________ SHEET NO. ________ OF ________

MH	ITEM	UNIT	QUANTITY	UNIT COST						TOTAL COST					
				L	PM	M&S	EO	P&E	T	LABOR	PM/SC	M&S	EO	P&E	TOTAL
	TOTAL DIRECTS ENGINEERING INDIRECTS & OVERHEAD														
	CONTINGENCIES GENERAL OVERHEAD & PROFIT														

The estimator will be responsible for these sheets. Table 5-3 is a summary of the accounts 0100 to 9800. Table 5-4 is listing for the equipment accounts 40xx. Table 5-5 is an example of the 0700 piling sheet and Table 5-6 is a simple overall summary sheet. Space is provided on each of these sheets to provide additional accounts if deemed necessary by the engineer or estimator. The type of quantities are also shown and the project manager should insist on each discipline quantifying his estimate to achieve estimated unit cost.

ESTIMATE. The first estimate produced in the early phase of design will be an order of magnitude estimate. It is best quantified in terms other than dollars or manhours, although the engineering disciplines do not like to be tied to a quantity, it should be required anyway. The basis for manhours and dollars should be given. Electrical estimates may be based on horsepower. Other examples are total tons of structural steel, or total cubic yards of concrete. These quantities are essential when substantiating estimate overruns later. If we can explain increases in cost because of increases in quantities, the bitter pill is not too hard to swallow.

Some details of estimating will now be considered using the civil, concrete and structural department for a specific instance.

For this department the information required for an estimate is

1. Description of project
2. Local building codes
3. Soil report, test borings, and test piles
4. General plant specifications
5. Civil, concrete, and structural specifications
6. Equipment lists
7. Plant plot plan

·8. Area equipment arrangement plans
·9. Plant firewater description and arrangement plans
·10. Process sewer description and arrangement plans
·11. Description and location of any facilities located outside of plant property
·12. Source of supply and cost of materials such as piling, fill, sand, rebar, form lumber, fabricated structural steel, and pea gravel
·13. Plot plan showing underground electrical requirements.

For this department the estimating responsibilities are

·1. Make labor and material take-off and price out all civil, concrete, and structural steel items
·2. Make topographical drawings for presentation with turnkey price
·3. Make surfaced area drawings for presentation with turnkey price
·4. Prepare write-up for civil, concrete, and structural work for presentation with turnkey price
·5. Make engineering estimate for civil, concrete, and structural steel requirements
·6. The civil, concrete, and structural steel department will be responsible for preparing estimates on prepared sheets for all areas as noted for the following functions and activities

FUNCTION	ACTIVITIES
0100	All activities
0200	All activities
0300	All activities
0700	All activities
0800	All activities Note: Insulation Dept. will pick up fireproofing for all items of equipment
0900	All activities Note: Cost of ladders and platforms for vessels on activities 0940, 0941, and 0942 will be prepared by Vessel Dept.

·7. Concrete quantities are to include overpours and seal slabs
·8. Include scaffolding for civil, concrete, and structural requirements
·9. Labor rates to be obtained from the Estimating Dept.

Check lists are used throughout the organization to ensure the inclusion of all considerations.

Accounts are kept in considerable detail, and the definition of accounts for 0700 PILING are

0700 PILING. All cost of driving piles. The quantity will be linear feet of piling driven
0710 Sheet piling. All costs of driving sheet pile
0711 Permanent sheet pile. All costs of driving and straightening permanent sheet pile
0720 Untreated wood piles. Includes all cost of driving untreated wood bearing and friction piles. Concrete caps will be included as a sub-feature under the concrete activity
0721 Drive. Driving and reseating heaved piles
0722 Load test. All cost of load test piles including special apparatus for testing
0723 Cut-off. Includes all costs of cutting off piles
0724 Steel cap. All costs of driving steel pile caps. Concrete will be included under the concrete account
0730 Treated wood piles. All cost of driving treated wood bearing and friction piles. Concrete caps will be included as a sub-feature under the concrete activity

BUDGET COMPLETION. The completion of the budget should be scheduled with as much devotion as for any other part of the project. The budget can be completed in the time allowed only if each member of the team is required to fulfill his obligations.

CONSTRUCTION. When the budget is completed the emphasis switches from engineering to construction. The construction project managers organization will appear as

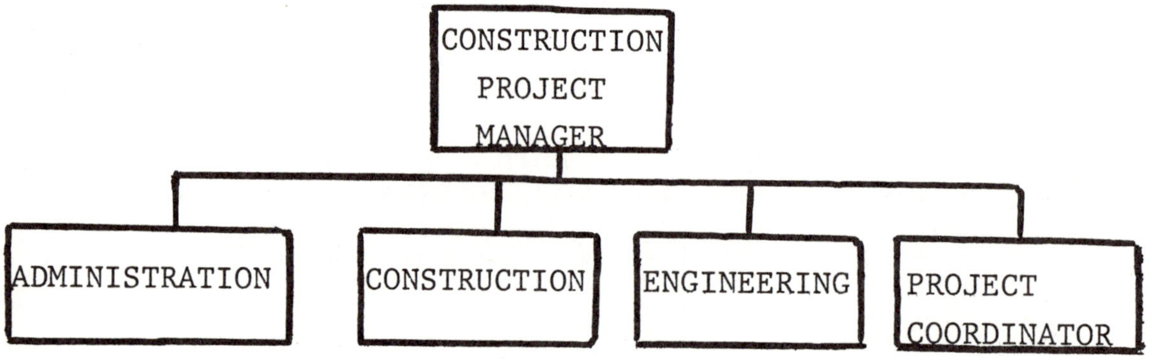

The project coordinator will handle management information systems like cost, estimating, and scheduling. Some are adding material control to these responsibilities. In any case it is

not wise to have split responsibility between cost, and estimating-scheduling.

VARIANCE. Variance reporting is a system which emphasizes cost and schedule control during all stages of design and construction. The purpose is to report all deviations from the budget to the client and to the company. The project cost engineer is responsible for coordinating the variance reporting system with engineering and construction. Each individual price change, quantity overrun, or scope change does not in itself become apparent, but the cumulative effect over extended time can become enormous. Table 5-7 shows a form for a project variance notice. Table 5-8 is a form for preparing a weekly variance graph.

CASH FLOW. Fig. 5-1 shows the cash flow month by month and the accumulated cash flow for a project. It also shows the labor cost and the accumulated labor cost. The accumulated curves are widely used and are also known as S curves. They can be prepared for labor cost, material cost, percent structural iron completion, percent completion and any aspect which is worth controlling. The curves are a powerful means of deciding when corrective action is necessary.

ESLUM. As mentioned the budget estimate should be complete before the construction operations move to the project site. The construction costs will be coded to budgeted items for comparison. Therefore a computer is used to compare the budget with the actual situation. A computer print out sheet called an ESLUM spread sheet is shown in Table 5-9. Note that the computer printout is in similar format to the preprinted estimating sheets. ESLUM is an acronym for EStimated Labor Unit Manhours, but it also includes materials.

Table 5-7

ORIGINATOR'S COPY

PROJECT VARIANCE NOTICE

PROJECT CHANGE ORDER NUMBER

	DATE	P.V.N. NO.	REV. NO.

TO:	B&R JOB NOS.	VARIANCE ORIGINATION ☐ HOME OFFICE ☐ FIELD
PROJECT		COST/SCHEDULE AREA

DRAWING, SPECIFICATION, EQUIPMENT NO., OR OTHER REFERENCE

SECTION 1.0 — REASON FOR AND NATURE OF VARIANCE

DESCRIPTION AND/OR SKETCH

2.0

ORIGINATION

ORIGINATOR	SUPERVISOR

SECTION 3.0

SECTION 3.0 TO BE COMPLETED WHEN INFORMATION IS AVAILABLE

EFFECT ON COST	EFFECT ON SCHEDULE	EFFECT ON CONTRACT

VARIANCE EVALUATION (By Control Team)

HOME OFFICE		MATERIALS $	SUB $	FIELD			TOTAL $
MANHOURS	AMOUNT $			MANHOURS	LABOR $	OTHER $	

SECTION 4.0

VARIANCE CLASSIFICATION

CLIENT	INTERNAL	ECONOMIC
☐ PROCESS MODIFICATION ☐ DESIGN MODIFICATION ☐ SPECIFICATION MODIFICATION ☐ IMPROVE OPERABILITY ☐ SAFETY CONSIDERATIONS ☐ OTHER ______ (SPECIFY)	☐ ESTIMATE ☐ DESIGN ☐ PROCESS ☐ OTHER ______ (SPECIFY)	☐ CHANGE IN MATERIAL COST ☐ CHANGE IN LABOR COST ☐ RESULTS OF PROCUREMENT ACTION ☐ FIELD CONDITIONS ☐ OTHER ______ (SPECIFY)

SECTION 5.0

	SCOPE CHANGE ONLY
☐ APPROVED ☐ APPROVED-REQUEST CHANGE ORDER - TO CLIENT ☐ DISAPPROVED ☐ ADDITIONAL INFORMATION REQUIRED ☐ REQUEST CONFERENCE ☐ OTHER (SPECIFY) ______	☐ DISAPPROVED ☐ PREPARE CHANGE ORDER REQUEST ESTIMATE
RESPONSIBLE MANAGER	CLIENT RESPONSIBLE MANAGER

Table 5-8

VARIANCE GRAPH

CLIENT: ______________________

PROJECT: C.E. WORKSHOP PROJECT

LOCATION: ______________________

PERIOD ENDING: 3/3/75

JOB NUMBERS: XX-XXXX

COST/SCHEDULE AREA: OVERALL

VARIANCE TO DATE

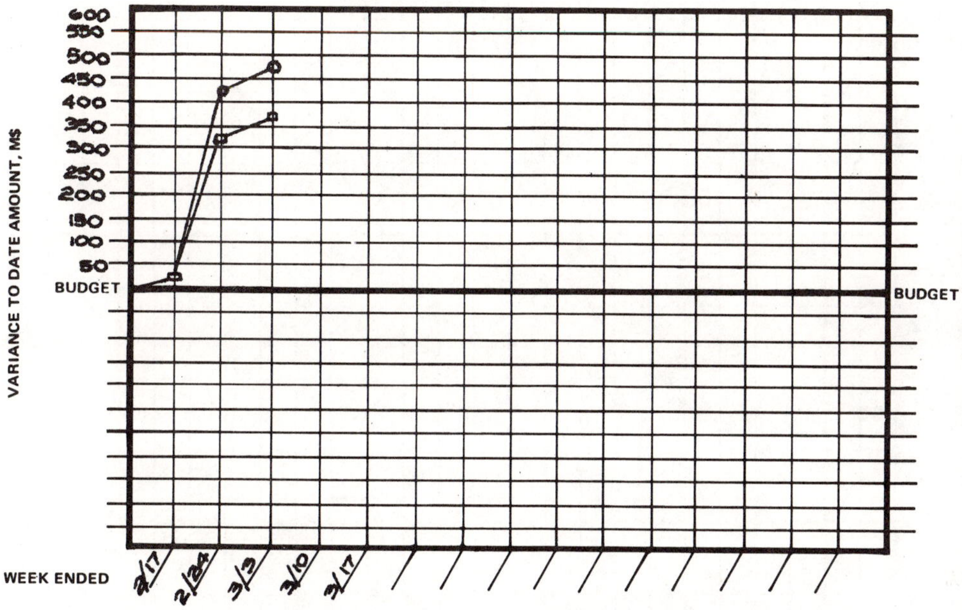

YEAR 1975

ALL COSTS IN THOUSANDS OF DOLLARS

SIGNIFICANT VARIANCES

□——□ Approved Project Change Orders

○——○ Project Variance Notices Plus Project Change Orders

Figure 5-1

THE CASH FLOW OF LABOR AND TOTAL COST

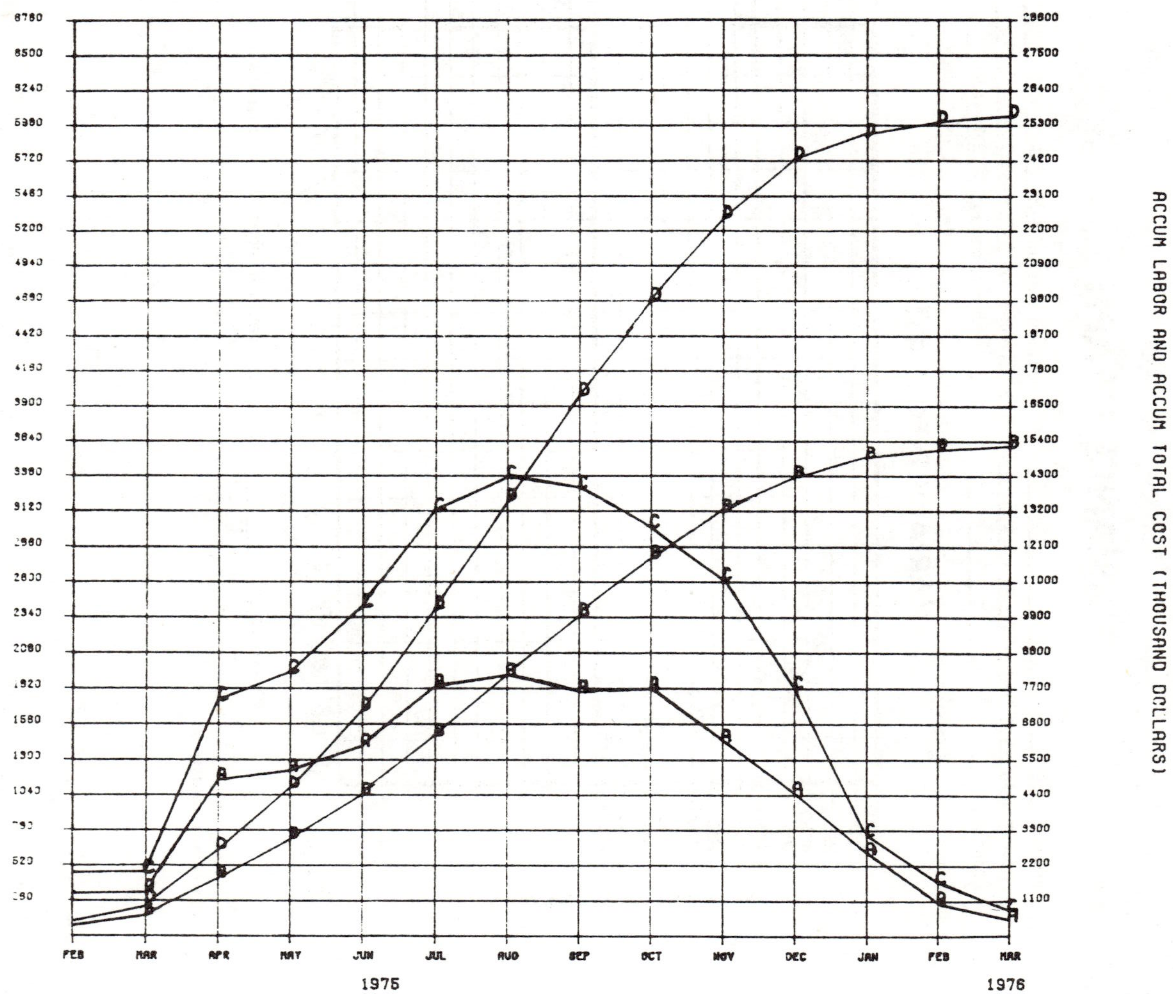

Legend:

Labor Cost A__A__A
Total Cost C__C__C
Accum Labor B__B__B
Accum Total D__D__D

This is a test job
Start Date February, 1975 Complete Date March, 1976.

Table 5-9

UNIT COST ACCOUNT	DESCRIPTION	QUANTITY	UNIT CODE	MANHOURS	LABOR (001)	PERM MTLS (002)	CONST MTL (003)	C/C	OTHER	UN/MH	TOTAL
031404036	STACKS			240	605				31,950		32,555
031404051	SCREENING EQUIPMENT										
40 5101	SCREEN, DBL DECK CW MOTORS & DRIVES	1	EA	300	755			109	9,100	300.000	9,855
031404051	SCREENING EQUIPMENT			300	755				9,100		9,855
031404058	DUST COLLECTORS										
40 5801	DUST COLLECTION UNITS FOR CALCINED LIME	5	EA	320	806			109	7,000	64.000	7,806
031404058	DUST COLLECTORS			320	806				7,000		7,806
031404059	OTHER										
40 5901	WET SCRUBBER FOR EXHAUST GAS	1	SET	3,060	7,710			109	51,000	60.000	58,710
031404059	OTHER			3,060	7,710				51,000		58,710
031404072	KILNS										
40 7201	ROTARY KILN 2.5M ID BY 60M LG	1	EA	9,910	24,980			109	518,000	910.000	542,980
40 7202	OIL BURNING EQUIPMENT CW COOLER TUBES&W/WAY	1	SET	4,890	12,315			109	151,200	890.000	163,515
031404072	KILNS			14,800	37,295				669,200		706,495
0314040	ENGINEERED EQUIPMENT			33,261	83,791	8,820	840		1,195,615		1,289,066
03140	LIME PREPARATION			177,355	306,613	752,527	96,924		1,240,865		2,396,999

COST INTEGRATED SCHEDULE REPORT is a type of management report that has both manhour cost and time displayed. It can be prepared on the Cal Comp Plotter from a computerized program and is shown in Fig. 5-2. The basic curves are the planned progress, the actual progress, and the earned value of work completed The presentation shows graphically the amount of overruns from the budget and the amount of time the project is behind schedule. The graph can reflect the total project as well as any elementary part of it. The area summary levels or activity summary levels could inform a project manager of amount of total slippage in both manhours and time. The amounts are bracketed in Fig. 5-2.

Table 5-10 shows an organization which pinpoints the basic responsibility necessary to achieve control. It has been mentioned that the project budget can be developed and controlled throughout the project through a project control manager. Also, that the control can be achieved by setting up and coordinating the efforts of the staff estimator, cost engineer, and scheduler. Finally,by establishing a separate area of responsibility at the construction level to coordinate these functions at the site, a function referred to as the "project coordinator" continuity is achieved to insure that overall project goals are realized.

The outline at the beginning of the chapter is an outline for cost control and reporting. The major topics and those judged to require the most emphasis have been discussed in the chapter ; but not all topics in the outline have been discussed. The outline will, however, serve as a checklist and reminder of what goes into good cost reporting and control.

Figure 5-2

EARNED VALUE STATUS GRAPH

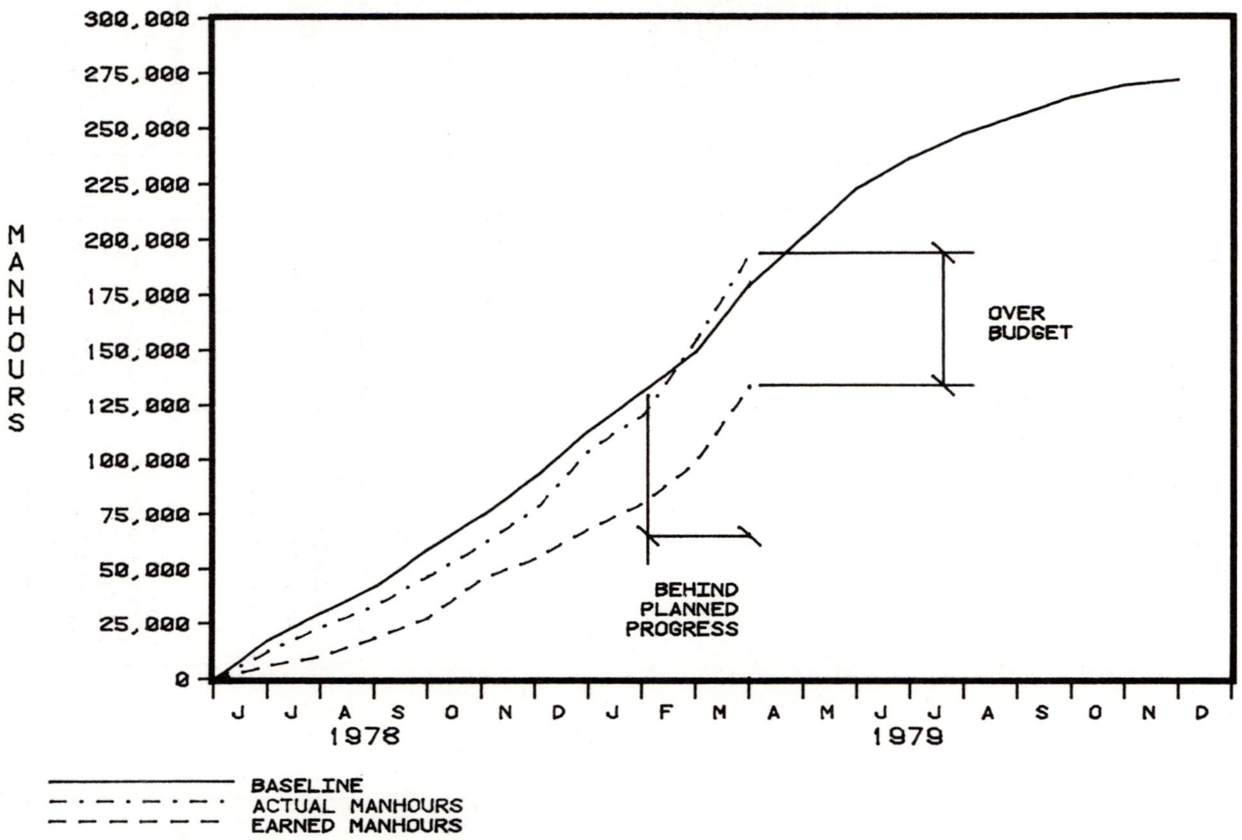

Table 5-10

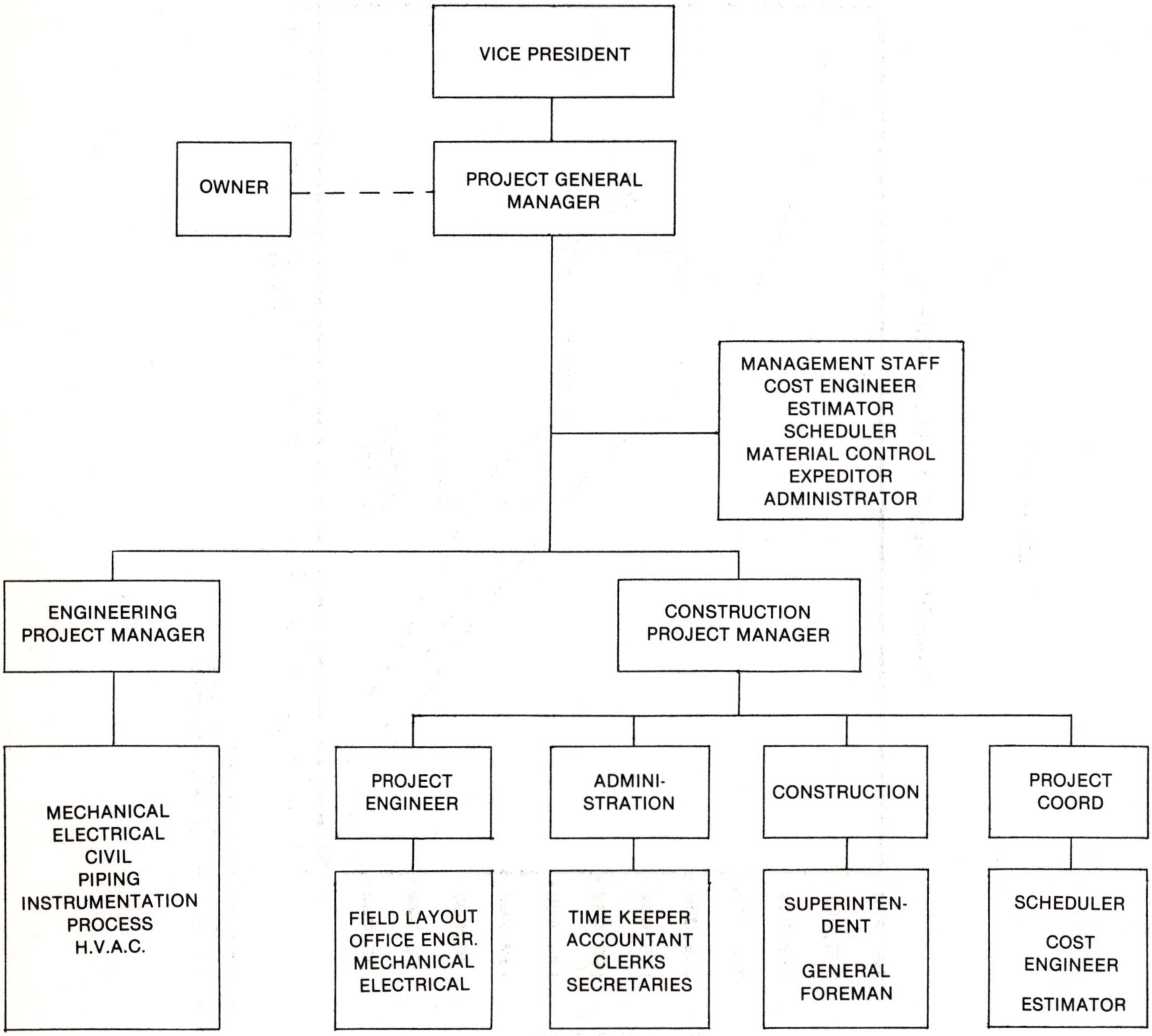
VICE PRESIDENT
OWNER
PROJECT GENERAL MANAGER
MANAGEMENT STAFF
COST ENGINEER
ESTIMATOR
SCHEDULER
MATERIAL CONTROL
EXPEDITOR
ADMINISTRATOR
ENGINEERING PROJECT MANAGER
CONSTRUCTION PROJECT MANAGER
MECHANICAL
ELECTRICAL
CIVIL
PIPING
INSTRUMENTATION
PROCESS
H.V.A.C.
PROJECT ENGINEER
FIELD LAYOUT
OFFICE ENGR.
MECHANICAL
ELECTRICAL
ADMINI-STRATION
TIME KEEPER
ACCOUNTANT
CLERKS
SECRETARIES
CONSTRUCTION
SUPERINTEN-DENT
GENERAL FOREMAN
PROJECT COORD
SCHEDULER
COST ENGINEER
ESTIMATOR

6. PROJECT MANAGEMENT AND SCHEDULE CONTROL

R. L. Dodds
Project Execution Planners, Houston, Texas

OUTLINE for this chapter follows:

I. Introduction
II. Project management, organization, and execution conflict
  A. Functional, projectized, and matrix organizations
  B. Organizational continuum
  C. Matrix organization and budget
  D. Conflict in project execution
III. Project management and control
  A. The real function of planning-setting milestones
  B. Time phasing the plan--scheduling--engineering, procurement, and construction
  C. Schedule and cost control--progress monitoring
  D. Cost engineering--field cost reporting, project cost reporting, and real cost control
IV. PERT, CPM and Precedence Comuterized Scheduling
  A. PERT, PERT/COST, and CPM defined
  B. Sample CPM problem
  C. CPM terms and definition
  D. Precedence--method and simplicity
  E. Commercial computer and schedule control programs

INTRODUCTION. The chapter covers project management and schedule control for the engineering and construction of major capital projects. It covers project organization, conflict in execution of projects, planning, time scheduling, cost control, and project monitoring. It includes CPM, Critical Path Method, and Precedence Computerized Scheduling Method. It lists some various sources that are available for scheduling systems.

PROJECT MANAGEMENT, ORGANIZATION, AND EXECUTION CONFLICT. Figures 6-1, 6-2, and 6-3 show the three types of organization--functional organization, projectized organization, and matrix organization. Fig. 6-1, functional organization, is the standard pyramid type with the organization broken down into functional groups or departments each reporting directly to the boss. In this arrangement engineering, for example, would report directly to the boss. In Fig. 6-2, projectized organization, the organization is by projects. In this arrangement, engineering, for example, would report to a project manager. Fig. 3, matrix organization, is a combination of the two previous organizations. In this arrangement engineering, for example, reports directly to the boss and to various project coordinators. Engineering would have teams working on various projects.

Small projects and projects in the range of up to 50 to 100 Mega$ will generally fall in the matrix organization, while projects above 100 Mega$ will fall in the projectized organization. The functional organization occurs chiefly where there are no projects, like a manufacturing operation.

Fig. 6-4 shows the matrix organization relationship. It defines the general manager, manager of projects, the functional manager and the various responsibilities.

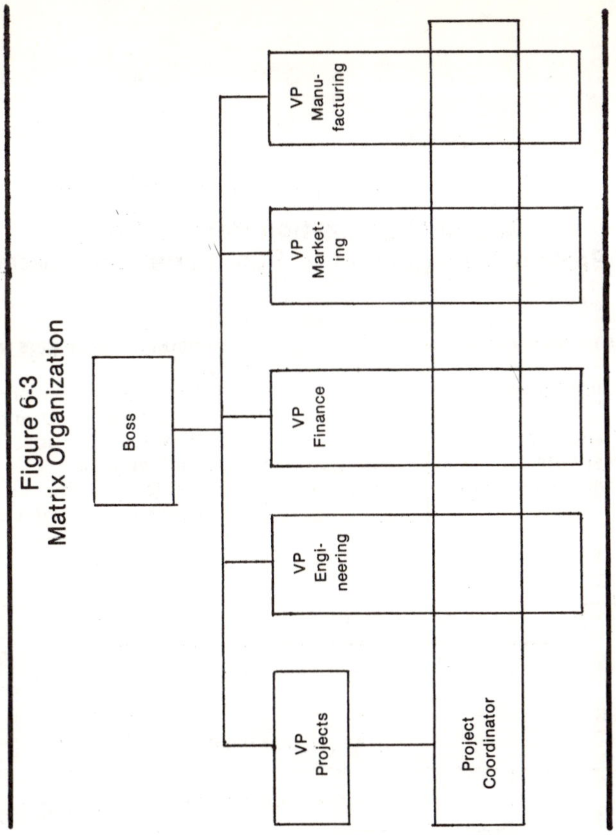

Figure 6-3
Matrix Organization

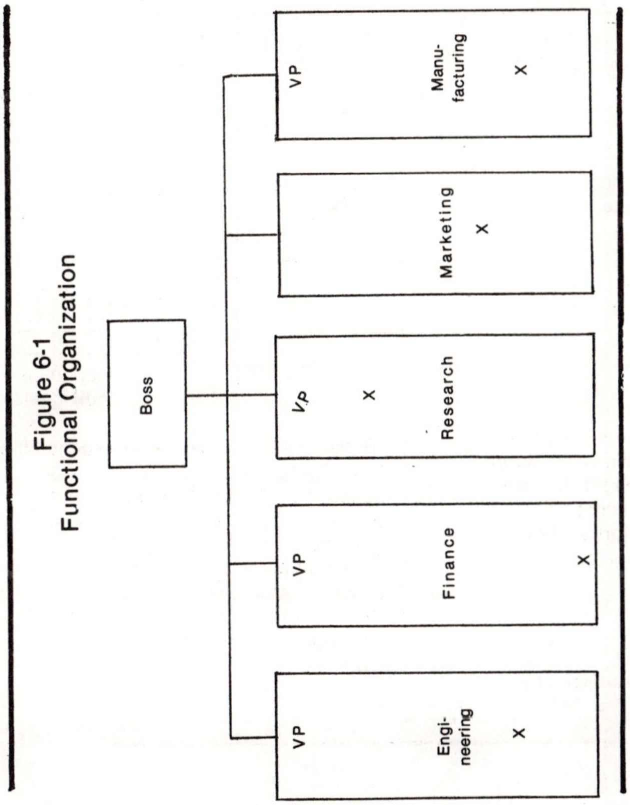

Figure 6-1
Functional Organization

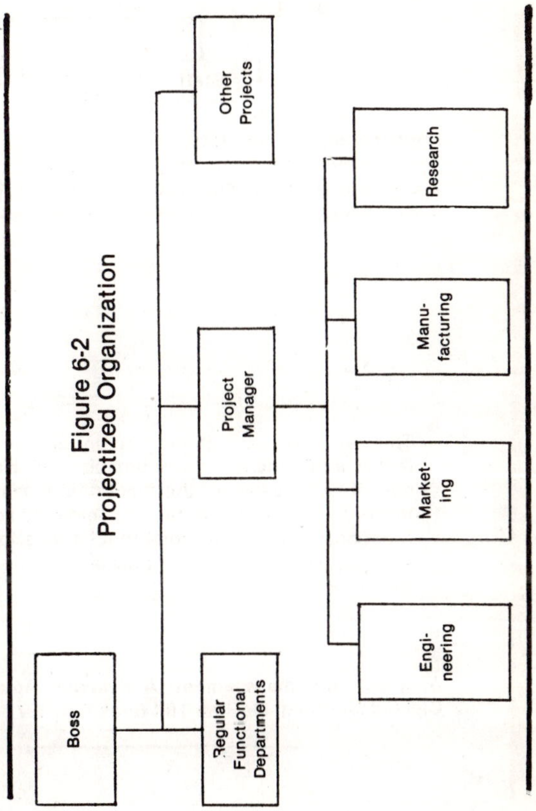

Figure 6-2
Projectized Organization

Reprinted by permission of The Project Management Institute, P.O. Box 43, Drexel Hill, PA 19026

Figure 6-4
Matrix Organization Relationship of Project Management to Functional Management

The Manager of Projects is Responsible for:
1. Directing and evaluating project manager activity.
2. Planning, proposing, and implementing project management policy.
3. Assuring project compliance with commitments.

Functional Managers are Responsible for:
1. Accomplishing work package tasks on schedule and within budget.
2. Providing functional policy and procedural guidance.
3. Providing adequately skilled staff.
4. Maintaining technical excellence.

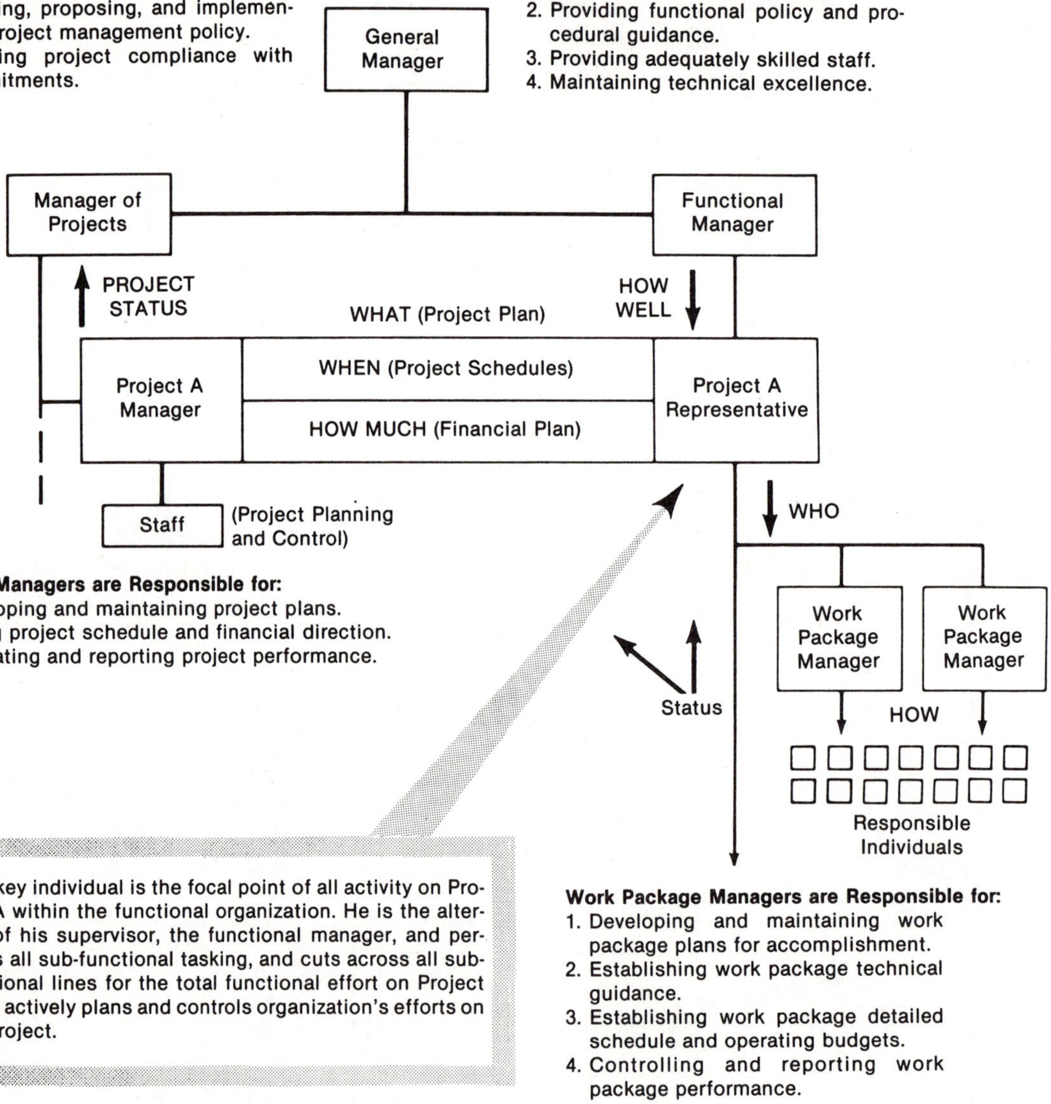

Project Managers are Responsible for:
1. Developing and maintaining project plans.
2. Giving project schedule and financial direction.
3. Evaluating and reporting project performance.

Work Package Managers are Responsible for:
1. Developing and maintaining work package plans for accomplishment.
2. Establishing work package technical guidance.
3. Establishing work package detailed schedule and operating budgets.
4. Controlling and reporting work package performance.

Adapted from **Management: A Systems Approach,** by Cleland and King, McGraw-Hill Book Co., 1972, p. 347.

Fig. 6-5 shows how the three major organizational forms appear as a continuum from 100% functional at the left side to 100% projectized at the right end, with the matrix form in between. The diagram breaks down the staff into two groups--the staff in functional departments and the staff on project teams. The diagram shows, too, that a weak matrix will have a part-time coordinator, moving to a full time coordinator as it becomes stronger and on to a full time project manager as it becomes still stronger. As projects get larger and larger the organization continuum moves to the right.

Fig. 6-6 shows the situation for a matrix organization and matrix budget. For an example the Project Manager A will add horizontally across the functional departments to get his total budget. What usually happens is that each functional department becomes overcommitted. As a result, conflicts will arise and politics enters. If the project manager does not control the budget, he must bargain with the functional departments. A matrix budget gives the project manager authority for resources from the functional departments. The project manager for Project A must be a diplomat to get the work done. Actually, the budget requires careful planning. The following items will add to the success of the matrix budget if it is to survive

- Top management must define responsibilities and authority for the project manager
- The project coordinator must anticipate conflicts
- Take positive steps to develop teamwork
- Use documentation to hold departments to their commitments
- Functional managers should review and sign all documents relating to plans and schedules for the project
- Avoid direct conflict with the department heads
- Project coordinator should limit himself to "what" is to be done not "how"
- Each project is a new effort. Careful and continuous planning is necessary.

Figure 6-5
Organizational Continuum

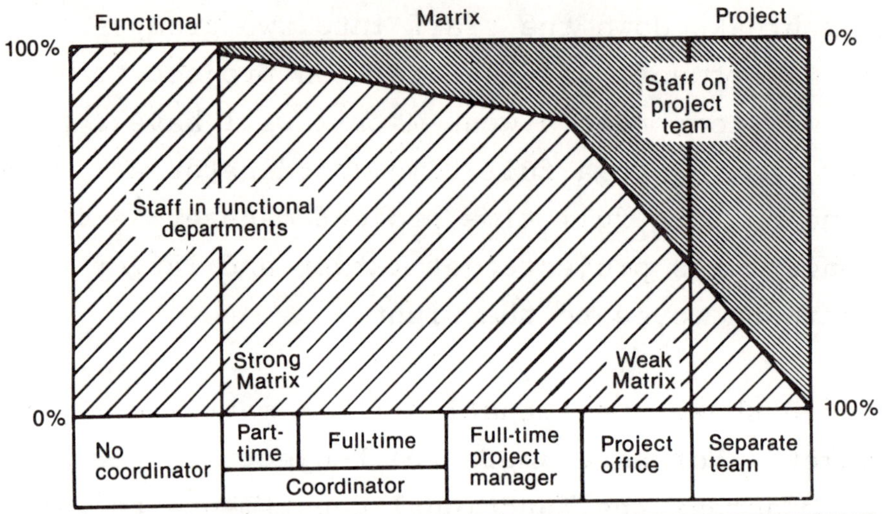

Figure 6-6
Matrix Organization and Matrix Budget

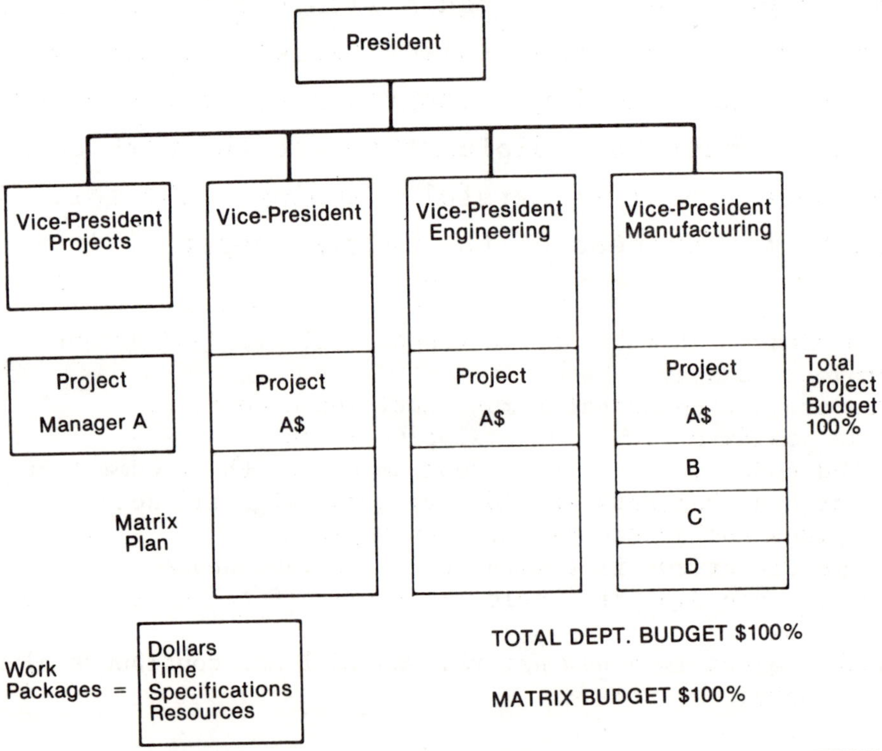

PROJECT MANAGEMENT AND CONTROL. We are led next to planning. Although many people have the ability to schedule, only a few are expert planners. The secret to a good plan is to put the whole project on a single sheet of paper as is shown in Fig. 6-7. It shows each phase of the project from contract award on April 13, 1976 to the mechanical completion on September 30, 1977. The Figure should be studied to comprehend the depth of information and timing presented on this single sheet.

Another example of a master plan is given in Fig. 6-8. It shows a Phase 1 and Phase 2 shutdowns in the shaded areas. The shutdowns were important and critical as they meant a loss to the refinery while the modernization was proceeding.

A very simple plan for a small project is shown in the form of a bar chart in Fig. 6-9.

The next step is to arrange a schedule from the plan. Key elements are the major milestones and Figs. 6-7, 6-8, 6-9 show clearly the major milestones like contract award dates, mechanical completion, start of shutdown and the like. The schedule must meet the major milestones to achieve the objectives set out in the plan. A first step in preparing the schedule from the plan is to prepare a plot plan as shown in Fig. 6-10. A plot plan shows the times for deliveries of various pieces of equipment and things like fabricated piping. Fig. 6-10 shows that the reactor steel is due in June.

On large jobs a more elaborate plot is required as shown in Fig. 6-11 which is for the fabricated steel. An overall plot plan shows multiple areas and work in blocks. It shows each area, delivery dates, vendors, etc. There will be individual plot plans for foundations, steel deliveries, piping de-

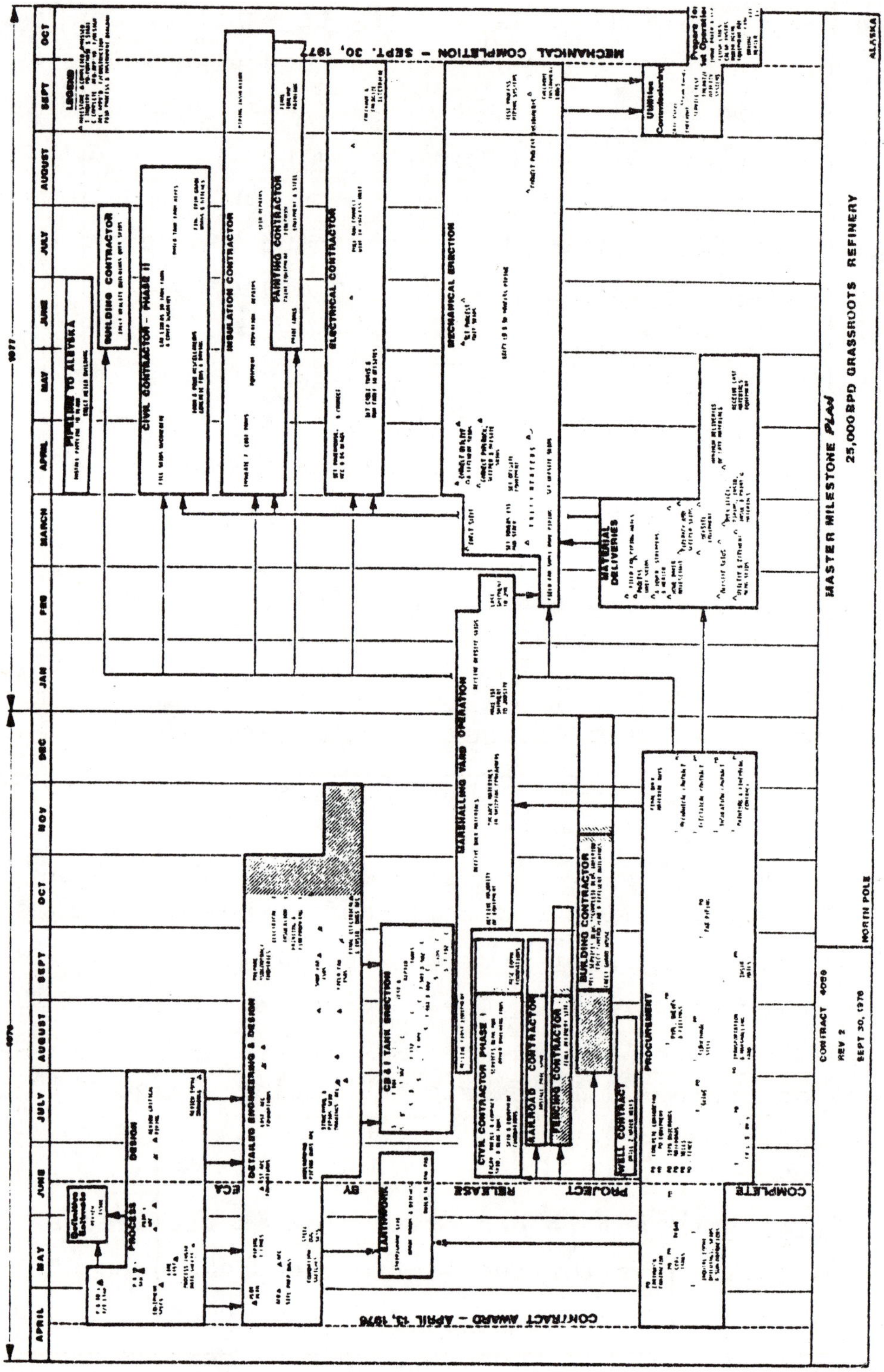

Fig. 6-7
Master milestone plan

Fig. 6-8
Project master plan

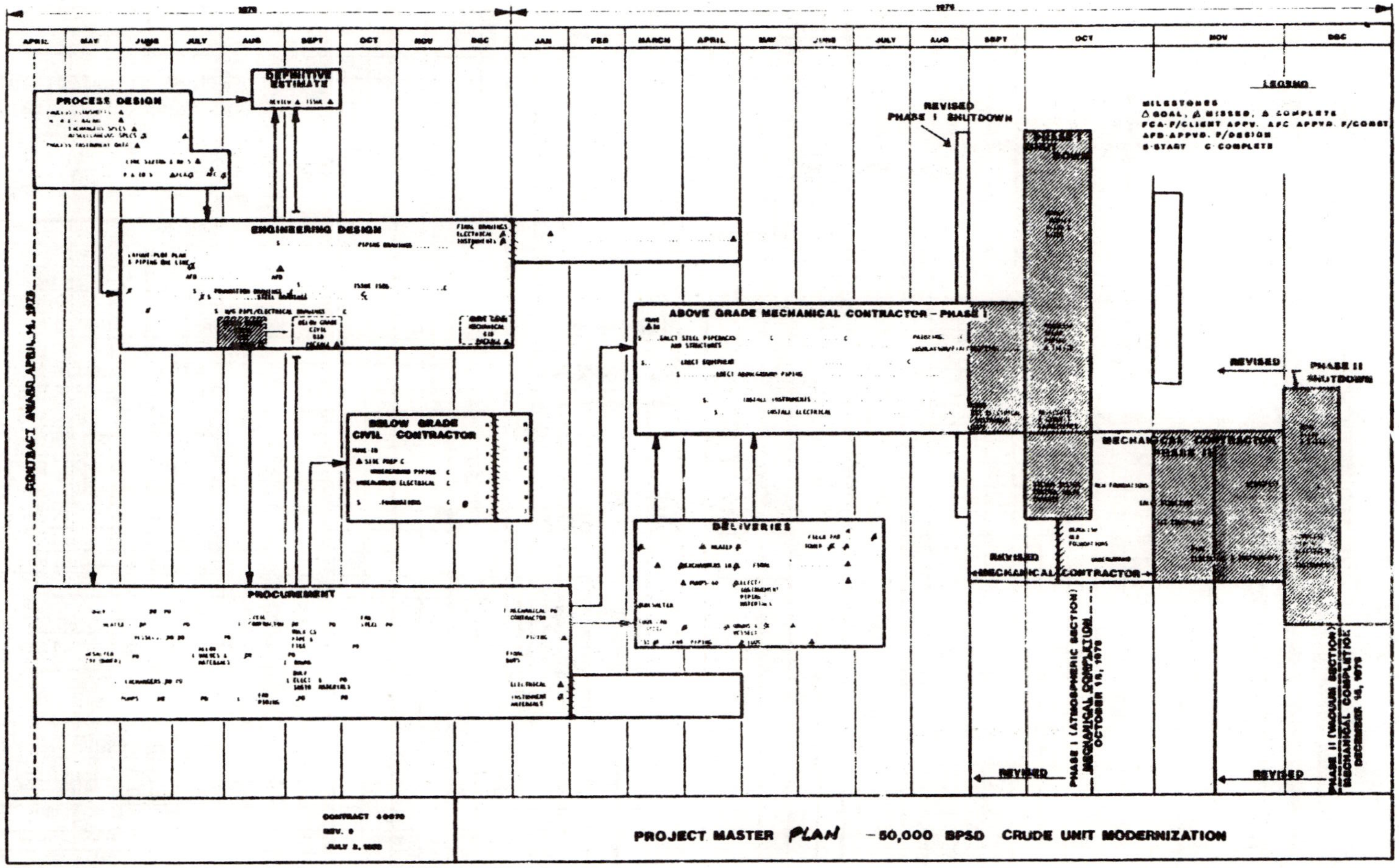

Fig. 6-9
Bar chart for project master plan

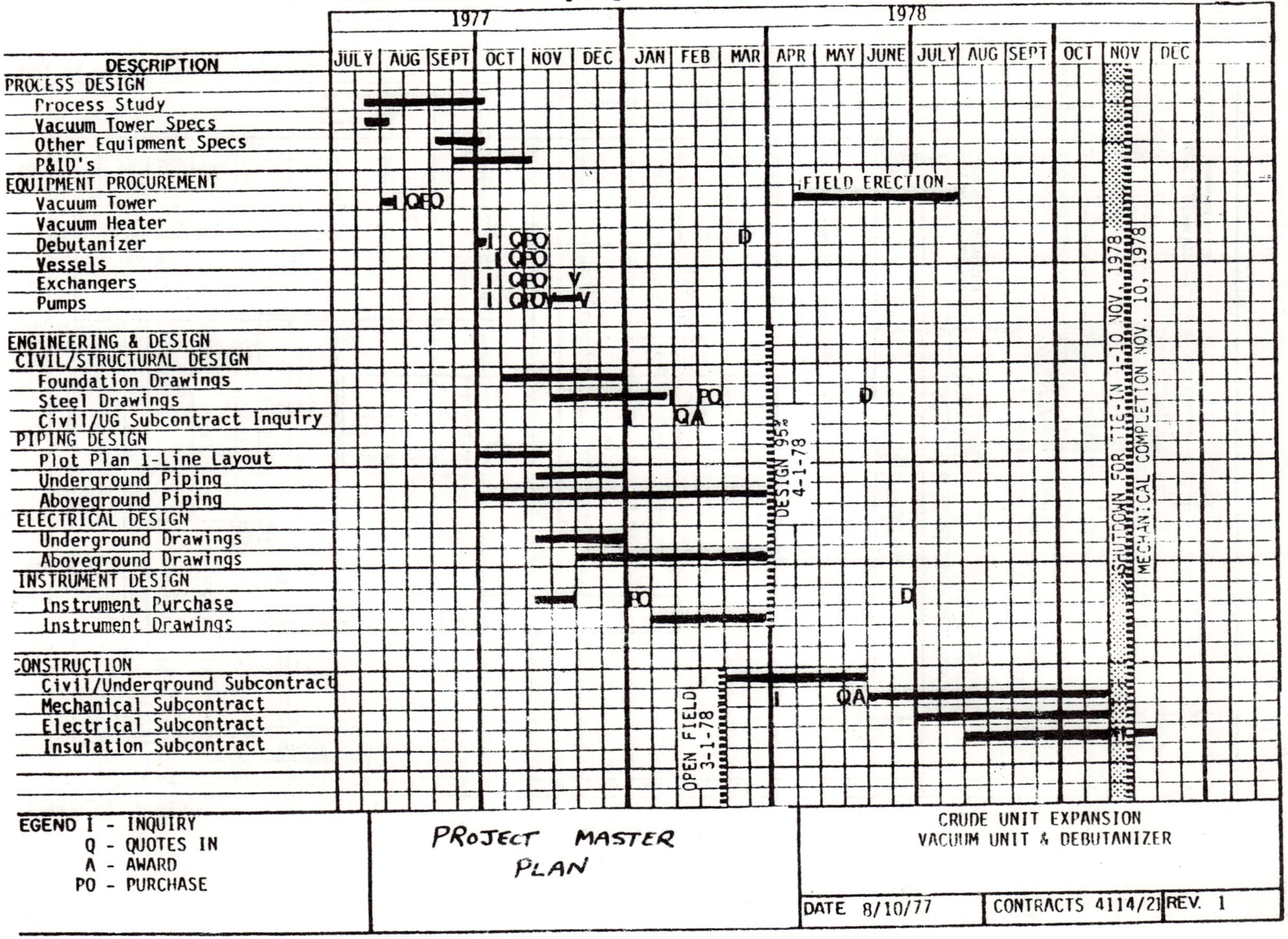

Fig. 6-10
Plot plan

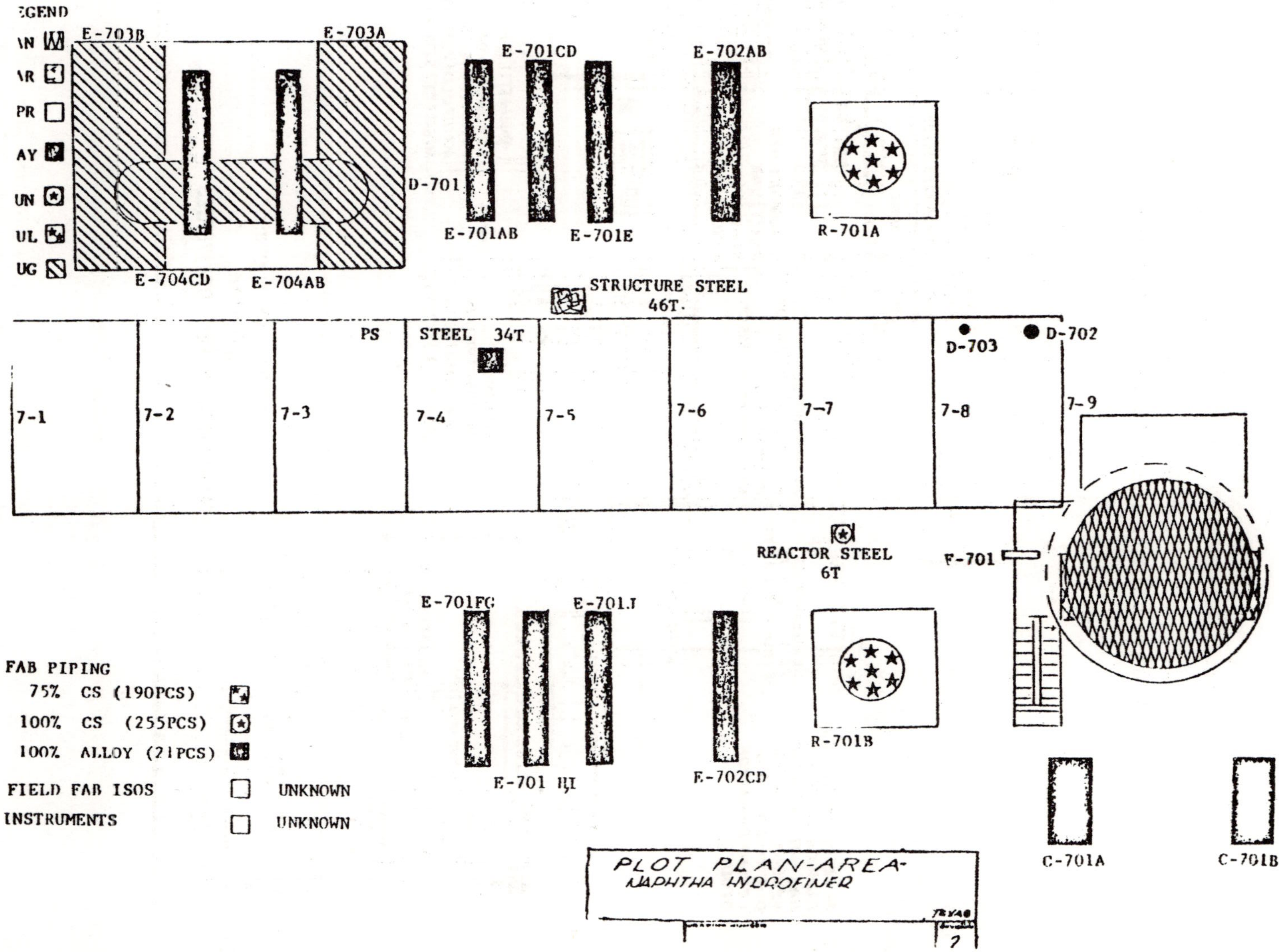

Fig. 6-11
Plot plan--fabricated steel deliveries

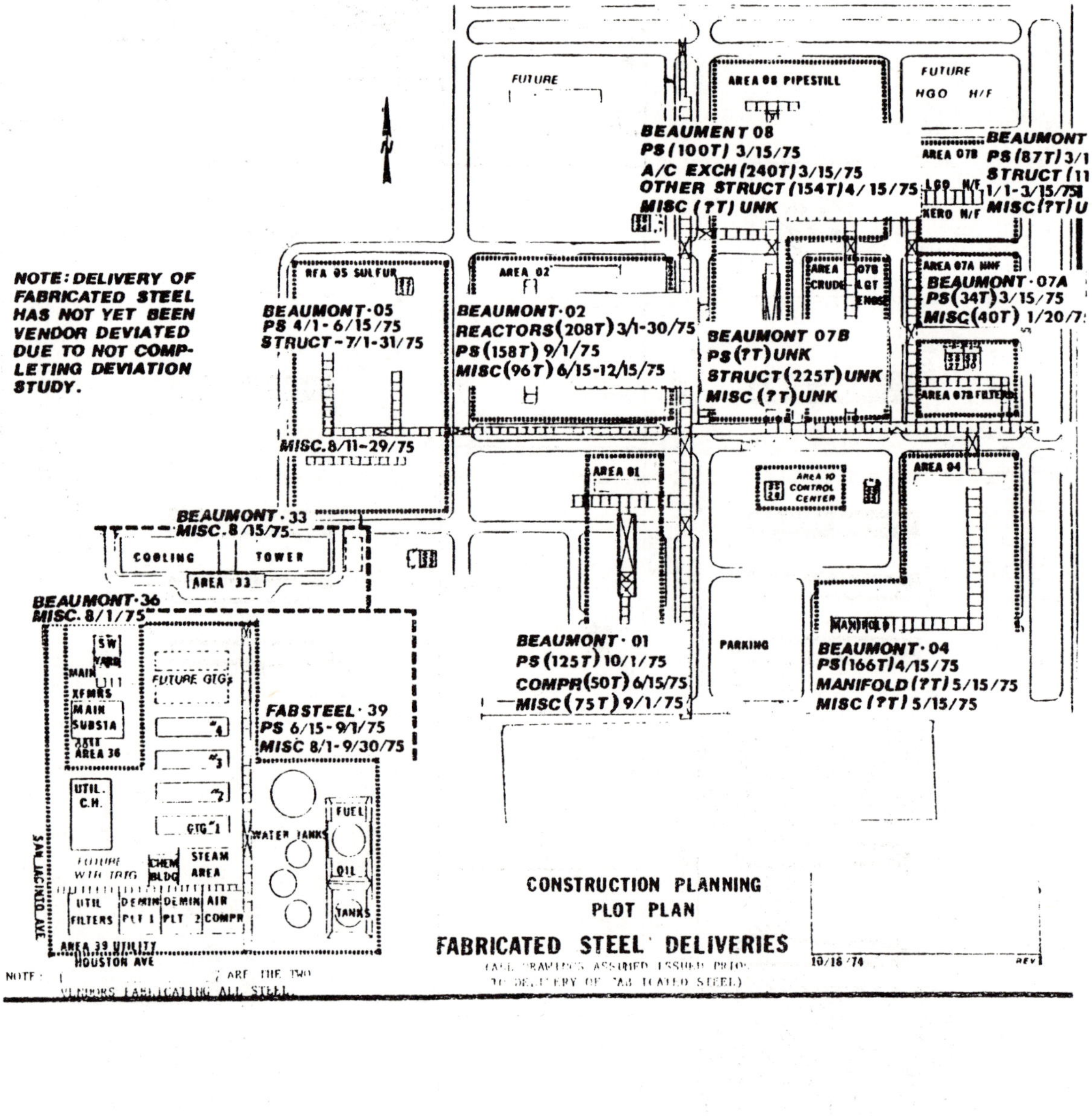

liveries, etc. One has to know delivery dates to do a good scheduling job.

Equally important is the planning for manpower like that shown in Fig. 6-12 for a moderate size project. Manpower requirements for a very large project are shown in Fig. 6-13. On large projects in can be difficult to satisfy the manpower requirements. For this particular job good planning and scheduling reduced the peak manpower requirement from 3800 to about 2500.

The end result of all planning is to come up with a time phased plan--called scheduling. Fig. 6-14 is a simple office design schedule shown as a bar chart, also called a Gantt chart.

Fig. 6-15 is a procurement schedule. It shows an equipment list and everything that must be done for equipment at each of the various milestones. The engineering status shows each of the milestone dates like process data, issue of specifications etc. Procurement is scheduled by scheduling each major milestone for each major piece of equipment. That can be an involved effort and the task should be computerized on a large project which might have up to two thousand pieces of major equipment to keep up with.

Fig. 6-16 shows a construction schedule. It is a weighted bar chart and the triangles show original delivery dates, revised delivery dates, and the actual arrival. Construction people like this form of chart.

Another example of a construction schedule is shown as a bar chart in Fig. 6-17 which also shows manhours. There is an interplay at D-701 since steel must be placed and a vessel must be set before beginning the piping effort.

Fig. 6-12
Anticipated construction manpower

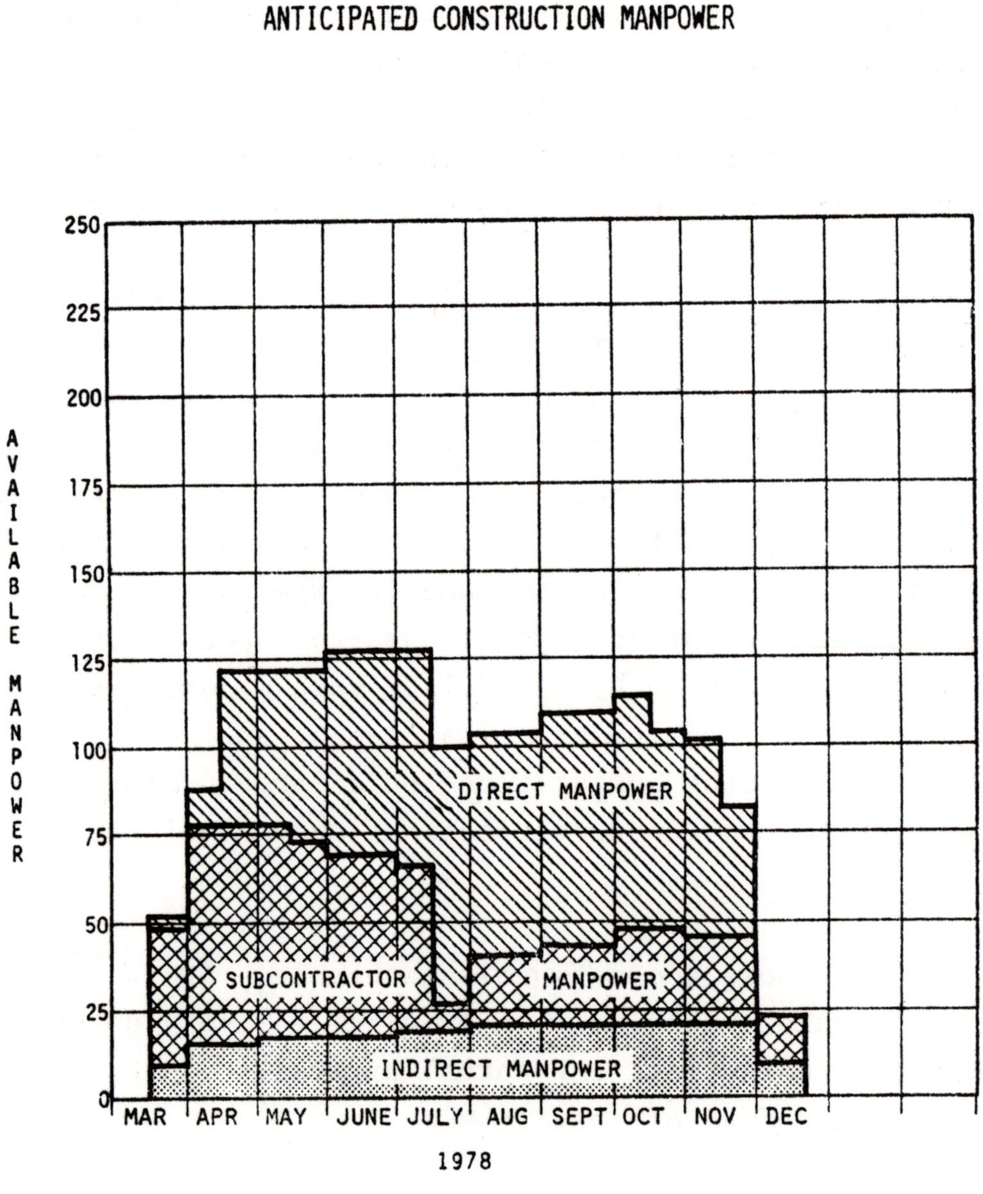

Fig. 6-13

Manpower for large project

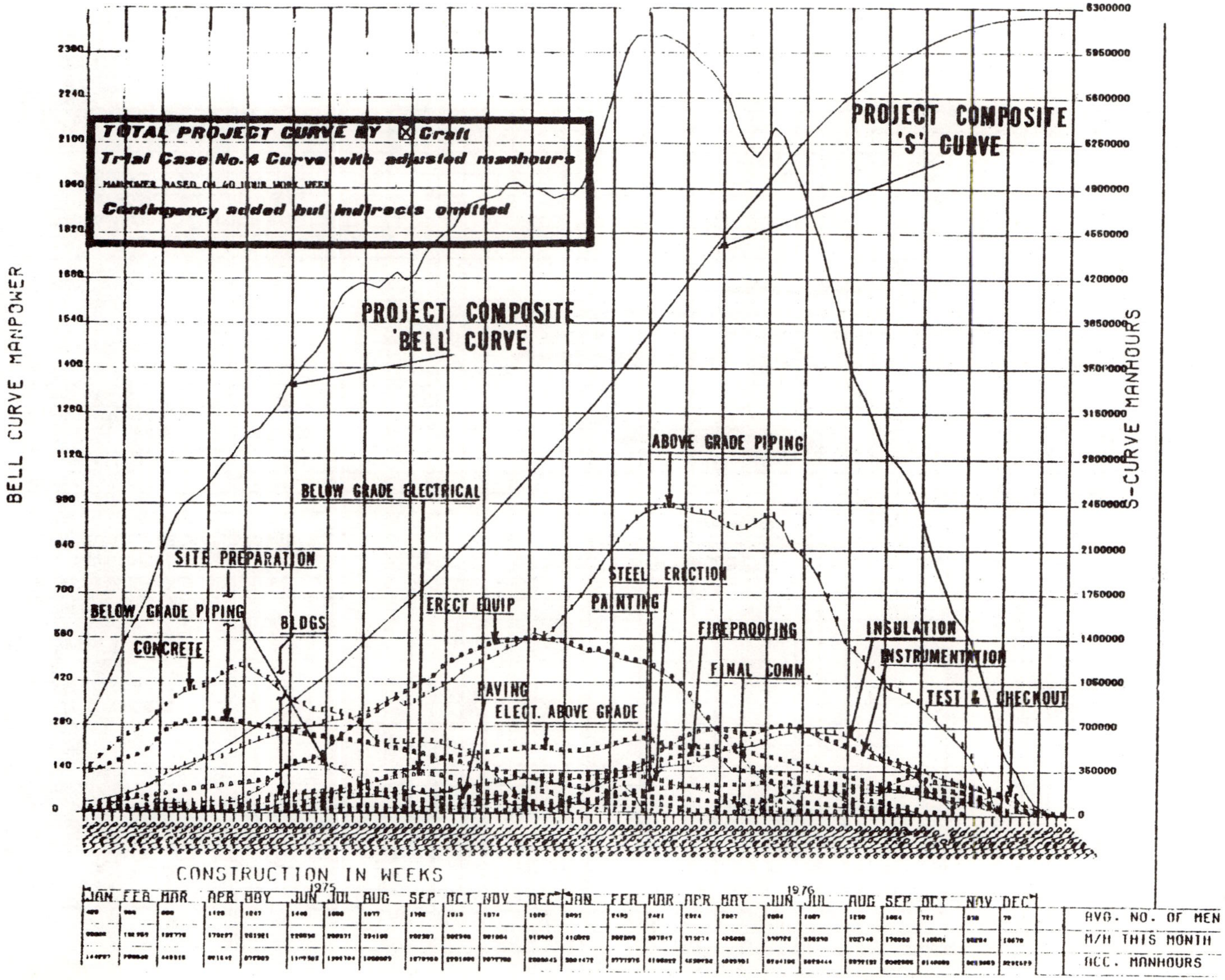

Fig. 6-14

Office designs schedule

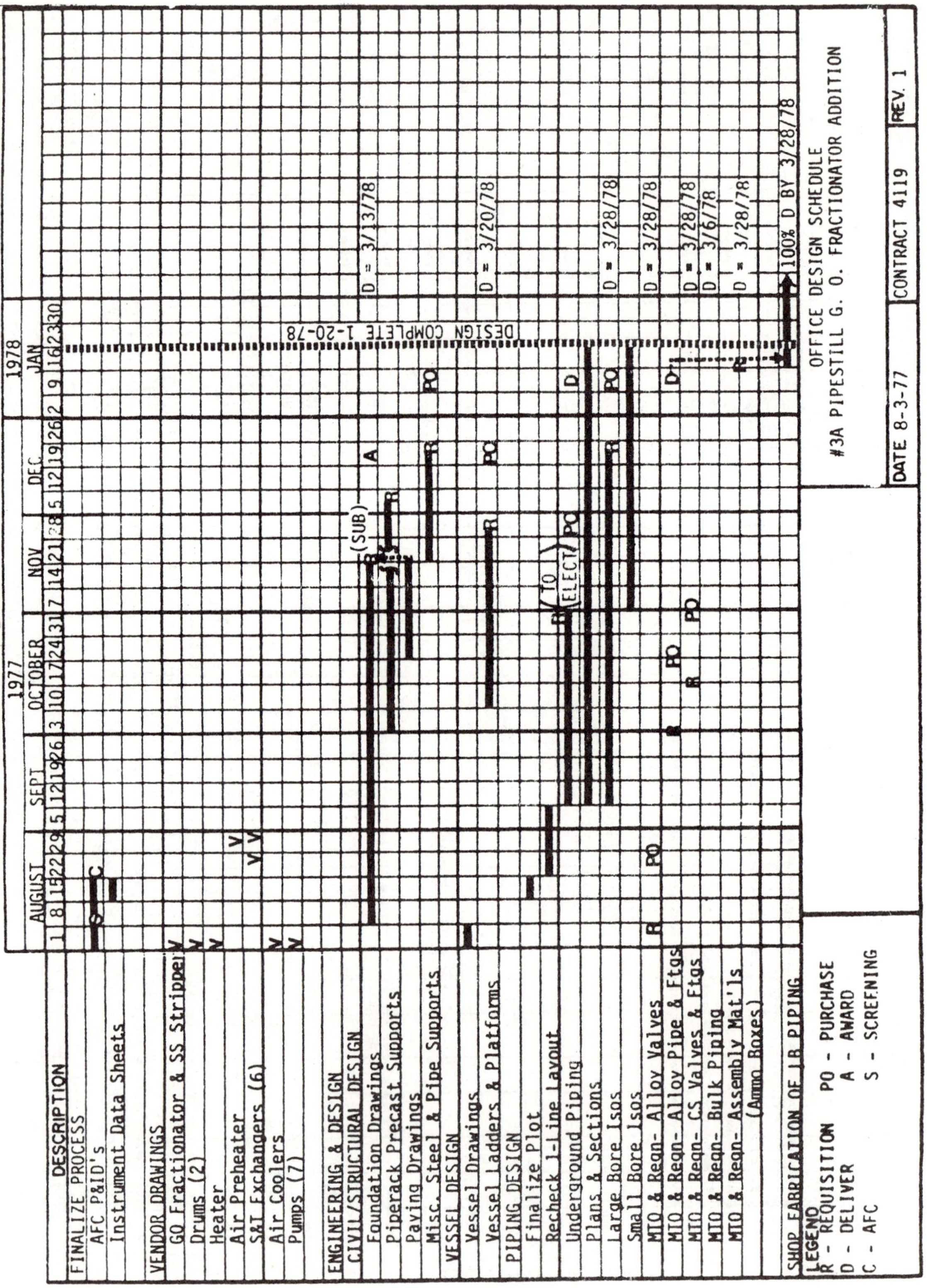

Fig. 6-15

MASTER EQUIPMENT LIST

CONTRACT 4109 | UNIT 25.1
CLIENT
COST CODE 1110 | CATEGORY Towers
BY STEVE | APPD
DATE 8-12-77 | PG 1 | OF 10 | REV 1

ITEM NO / COST CODE	DESCRIPTION AND SIZE	WEIGHT LBS	REQN NO	REQN REV	REQN DATE	P.O. NO REV	P.O. DATE	VENDOR / REMARKS
95-201 / 1111-01	CRUDE TOWER 19'-0" I.D.x148'-1" T/T 192'-0" O.A.L.	651,850	1111-1	0 / 1		0 / 1		A285C w/Monel & 410s Clad
1112-01	CRUDE TOWER TRAYS	48	1112-1	0 / 1	6/21	0 / 1		12 cr.
95-202 / 1111-02	HEAVY NAPHTHA STRIPPER* 5'-0" I.D.x23'-6" T/T	*63,675	1111-2	0 / 1	6/29 / 7/22	0 / 1		*Stacked Tower 91'-8" O.A.L. A285C
1112-02	HEAVY NAPHTHA STRIPPER TRAYS	6	1112-2	0 / 1	6/29	0 / 1		12 cr.
95-203 / 1111-02	NO. 2 FUEL OIL STRIPPER* 9'-6" I.D.x24'-0" T/T	*	1111-2	0 / 1	6/29 / 7/22	0 / 1		A285C w/410s Clad
1112-02	NO. 2 FUEL OIL STRIPPER TRAYS	6	1112-2	0 / 1	6/29	0 / 1		12 cr.
95-204 / 1111-02	LGO STRIPPER* 5'-0" I.D.x31'-6" T/T	*	1111-2	0 / 1	6/29 / 7/22	0 / 1		A285C w/410s Clad
1112-02	LGO STRIPPER TRAYS	6	1112-2	0 / 1	6/29	0 / 1		12 cr.
95-205 / 1111-03	DEPENTANIZER 7'-6"x9'-6" I.D.x146'-6" T/T O.A.L.	310,170		0 / 1		0 / 1		
1112-03	DEPENTANIZER TRAYS	64		0 / 1		0 / 1		12 cr.
95-206 / 1111-04	AMINE CONTACTOR 5'-0" I.D.x48'x6" T/T 56'-9" O.A.L.			0 / 1		0 / 1		A515GR70
1112-04	CONTACTOR PACKING AND INTERNALS			0 / 1		0 / 1		

EQUIPMENT STATUS

LEGEND
L = LETTER
S = SLIPPAGE
T = TELEPHONE
V = VISIT

SHIPMENT CODE
UP TO 50% | RECEIVED
50%-75% | C CANCELLED
100% | H HOLDING

DATE 8-12-77 | PG 1 | OF 10 | REV.

	SCHEDULED DATES: DWGS FOR APPR	DWGS CERT.	SHIP FROM SHOP	ACTUAL SHIPPING DATES	DATES REQ'D @ SITE VIA ETA @ SITE
REQD / ORIGINAL / CURRNT $ / CKD BY	9/1	2/1		SHIPMT CODE / FIRST / LAST	REQD / ETA / ETA
REQD / ORIGINAL / CURRNT $ / CKD BY	9/9	12/1		SHIPMT CODE / FIRST / LAST	REQD / ETA / ETA
REQD / ORIGINAL / CURRNT $ / CKD BY	10/12	12/15		SHIPMT CODE / FIRST / LAST	REQD / ETA / ETA
REQD / ORIGINAL / CURRNT $ / CKD BY	9/15	12/15		SHIPMT CODE / FIRST / LAST	REQD / ETA / ETA
REQD / ORIGINAL / CURRNT $ / CKD BY	10/12	12/15		SHIPMT CODE / FIRST / LAST	REQD / ETA / ETA
REQD / ORIGINAL / CURRNT $ / CKD BY	9/15	12/15		SHIPMT CODE / FIRST / LAST	REQD / ETA / ETA
REQD / ORIGINAL / CURRNT $ / CKD BY	10/12	12/15		SHIPMT CODE / FIRST / LAST	REQD / ETA / ETA
REQD / ORIGINAL / CURRNT $ / CKD BY	9/15	12/15		SHIPMT CODE / FIRST / LAST	REQD / ETA / ETA
REQD / ORIGINAL / CURRNT $ / CKD BY	10/16	12/15		SHIPMT CODE / FIRST / LAST	REQD / ETA / ETA
REQD / ORIGINAL / CURRNT $ / CKD BY	10/20	12/3		SHIPMT CODE / FIRST / LAST	REQD / ETA / ETA
REQD / ORIGINAL / CURRNT $ / CKD BY	10/20	12/31		SHIPMT CODE / FIRST / LAST	REQD / ETA / ETA
REQD / ORIGINAL / CURRNT $ / CKD BY				SHIPMT CODE / FIRST / LAST	REQD / ETA / ETA

ENGINEERING STATUS

DATE 8-12-77 | PG 1 | OF 10 | REV. 1

DATES	PROCESS DATA	ISSUE SPEC & REQN	INQUIRY ISSUED	QUOTES REC'D	BID TAB	CLIENT APP'D BID TAB	RELEASE FOR PO
SCH	4/27	5/25	5/26	6/24	7/20	7/29	8/1
ACT	4/27	5/25	5/26		7/21		
			INQ NO 1-1100				
SCH	6/10	6/21	6/22	7/15	7/27	8/10	8/11
ACT	6/10	6/21					
			INQ NO 2-1100				
SCH	7/15	7/21	7/22	8/19	8/24	9/8	9/9
ACT	7/15	7/23	7/25				
			INQ NO 4-1100				
SCH	6/24	6/24	6/29	7/22	7/29	8/11	8/12
ACT	6/20	6/30	6/30				
			INQ NO 3-1100				
SCH							
ACT							
		SEE	INQ NO H.N. STRIPPER				
SCH							
ACT							
		SEE	INQ NO H.N. STRIPPER TRAYS				
SCH							
ACT							
		SEE	INQ NO H.N. STRIPPER				
SCH							
ACT							
		SEE	INQ NO H.N. STRIPPER TRAYS				
SCH	7/12	7/27	7/28	8/26	8/31	9/11	9/12
ACT	7/12						
			INQ NO				
SCH	7/15	7/27	7/28	8/26	9/1	9/15	9/16
ACT	7/15						
			INQ NO				
SCH	7/12	7/27	7/28	8/26	9/1	9/15	9/16
ACT	7/12						
			INQ NO.				
SCH							
ACT							
			INQ NO.				

Fig. 6-16

Construction work sheet

△ ORIGINAL DELIVERY DATE
△ REVISED DELIVERY DATE
▲ DATE RECEIVED AT SITE
[ ] ACTIVITY START & FINISH

DATE APRIL 1, 19 77
PAGE 5 OF 5

CUSTOMER
LOCATION
CONTRACT No. 1086

CONSTRUCTION WORK SHEET

1977

DESCRIPTION	WT'D UNIT	WT'D UNIT COMPL	% COMPL	REMARKS
33 - [illegible] - OSBL				
EQUIPMENT SETTING				
PUMPS				
P-912 A/B	60	30	50	PUMPS MUST BE REPAIRED
P-320	30	5	50	PUMPS MUST BE REPAIRED
MIXERS/AGITATORS				
M-121	25	25	100	
PROCESS SKIDS				
[illegible]	[illegible]	[illegible]	100	
6	80	80	100	
13	[illegible]	32	81	
PIPE SKIDS				
101/102	39	39	100	
103/104	39	39	100	
105/106	39	39	100	
107/108	39	39	100	
109/110	39	39	100	
111/112	34	34	100	
113/114	39	39	100	
115/116	39	39	100	
117/118	97	97	100	
119/120	107	102	95	
121/122	116	116	100	
123/124	68	68	100	
125	20	20	100	
1A/1B	291	218	75	DAMAGED IN SHIPMENT
2A/2B	267	200	75	DAMAGED IN SHIPMENT
201/202	107	102	95	
203/204	60	60	100	
205/206	146	146	100	
207/208	146	139	95	
312/313	24	24	100	
314/404	21	21	100	
405/406	21	21	100	
407/408	21	21	100	
409/410	21	21	100	
STRUCTURAL STEEL				
LADDERS & PLATFORMS	493	463	95	
PIPE RACKS & SLEEPERS	360	517	95	
MISC. SUPPORTS	40	40	100	
PIPING				
SMALL BORE FABRICATION	2108	2108	100	
SMALL BORE ERECTION	7216	4690	65	
LARGE BORE ERECTION	15659	13150	84	
BRACKETS & HANGERS	3780	2369	63	
STEAM/HOT OIL TRACING	4772	954	20	
INSTRUMENTATION				
FIELD INSTRUMENTS	455	296	65	
TRANSMITTERS	482	305	66	
INSTRUMENT PANELS	55	[illegible]	[illegible]	

Schedule columns (1977): FEB 5, 12, 19, 26; MAR 5, 12, 19, 26; APRIL 2, 9, 16, 23, 30; MAY 7, 14, 21, 28; JUNE 4, 11, 18, 25; JULY 2, 9, 16, 23, 30; AUG 6, 13, 20, 27; SEPT 3, 10, 17, 24

Fig. 6-17
Simple bar chart construction schedule

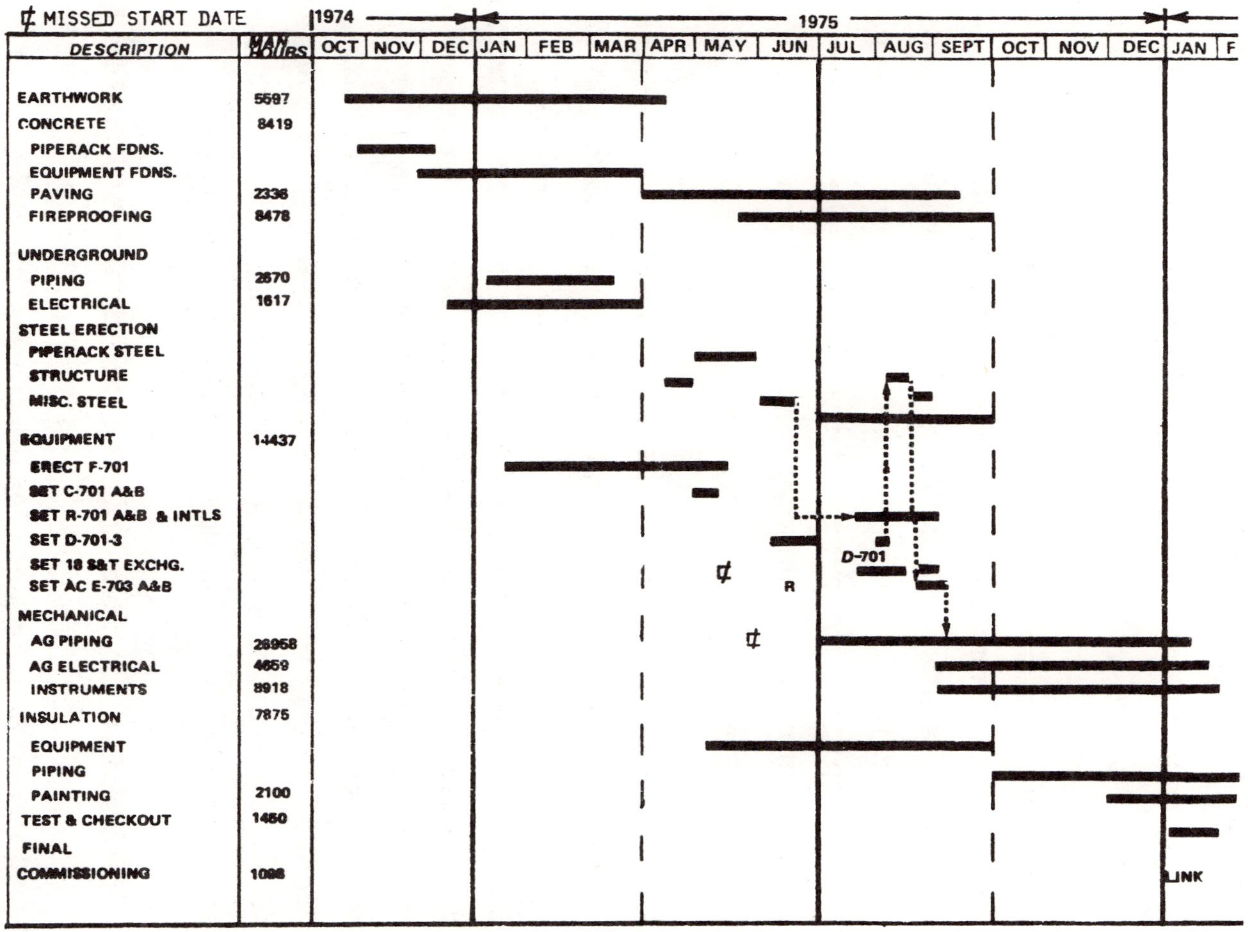

The Figures have shown the basic types of scheduling that will work for engineering, procurement, and construction scheduling.

Next we have to monitor the project to determine that the schedule is achieved. Figs. 6-18 through 6-23 are all examples of progress reports for engineering and procurement. Later, progress reports for construction will be shown. Fig. 6-18 is an office schedule performance summary which shows percent completion for the various disciplines in an engineering office, and the total project. It is an interim report showing what percent was completed on a particular date, also a comparison with two weeks earlier, and what was accomplished in the two weeks.

Fig. 6-19 shows a total project report at a certain date near the end of a project. The dotted line shows the schedule, the full line the actual progress, and the dot-dash line the % budget manhours spent. Fig. 6-20 shows a number of subsidiary curves for total design and the various design groups. Fig. 6-21 shows the curves for the engineering disciplines project engineering, process and flow sheets, procurement, and estimating and cost. All this should be augmented with computer printouts like Fig. 6-22 which is a status report for work done at a refinery and Fig. 6-23 another refinery project.

At this point we have covered progress monitoring for engineering and procurement of a contractor or an operating company which has a department for engineering and procurement. We now proceed to construction cost monitoring. The principles remain the same and Figs. 6-24 through 6-31 illustrate various progress displays. They are similar to Figs. 6-19 through Fig. 6-23. Fig. 6-27 features the field indirect costs which now have become a major part of large jobs.

Fig. 6-18

Office schedule performance summary

25,000 BPSD REFINERY

Progress as of 9/25/76

OFFICE SCHEDULE PERFORMANCE SUMMARY

	PROGRESS (% Complete)			SCHEDULE			MANHOURS		
	As Of 9/25	As Of 9/11	2 Week Gain	% Schd.	Ahead+ Behind-	2 Wks. Ago	% Mhrs.	Under - Over +	2 Wks Ago
Project/Equipment	50½%	48½%	2 %	50%	+½%	+1½%	50 %	-½%	-1 %
Process/Flow Sheets	91½%	90½%	1 %	94%	-2½%	-½%	97½%	+6 %	+4%
Purchasing	61½%	54½%	7 %	47%	+14½%	+10½%	42½%	-19 %	-15½%
MTO & Estimating	79 %	76½%	3½%	62%	+17 %	+18½%	47½%	-31½%	-31½%
Inspection/Expediting	24 %	21 %	3 %	19%	+5 %	+6 %	29 %	+5%	+2%
Cost & Scheduling	24½%	23½%	1 %	22%	+2½%	+2½%	14½%	-10%	-10 %
Design	85 %	79 %	6 %	82%	+3 %	+5 %	77%	-8 %	-9 %
Vessels	91 %	89 %	2 %	88%	+3 %	+7 %	66½%	-24½%	-25 %
Civil/Structural	98 %	98 %	0 %	87%	+11 %	+17 %	68 %	-30 %	-31½%
Piping	79½%	73½%	6 %	81 %	-½%	+½%	83%	+3½%	+1 %
Electrical	69 %	55½%	13½%	70%	-1 %	-6½%	80 %	+11%	+9 %
Instruments	86½%	76½%	10 %	78%	+8½%	+6½%	76 %	-10½%	-8½%
TOTAL PROJECT	63 %	59%	4 %	58%	+5%	+5%	55½%	-7½%	-8 %

Fig. 6-19
Office schedule progress curve--total project

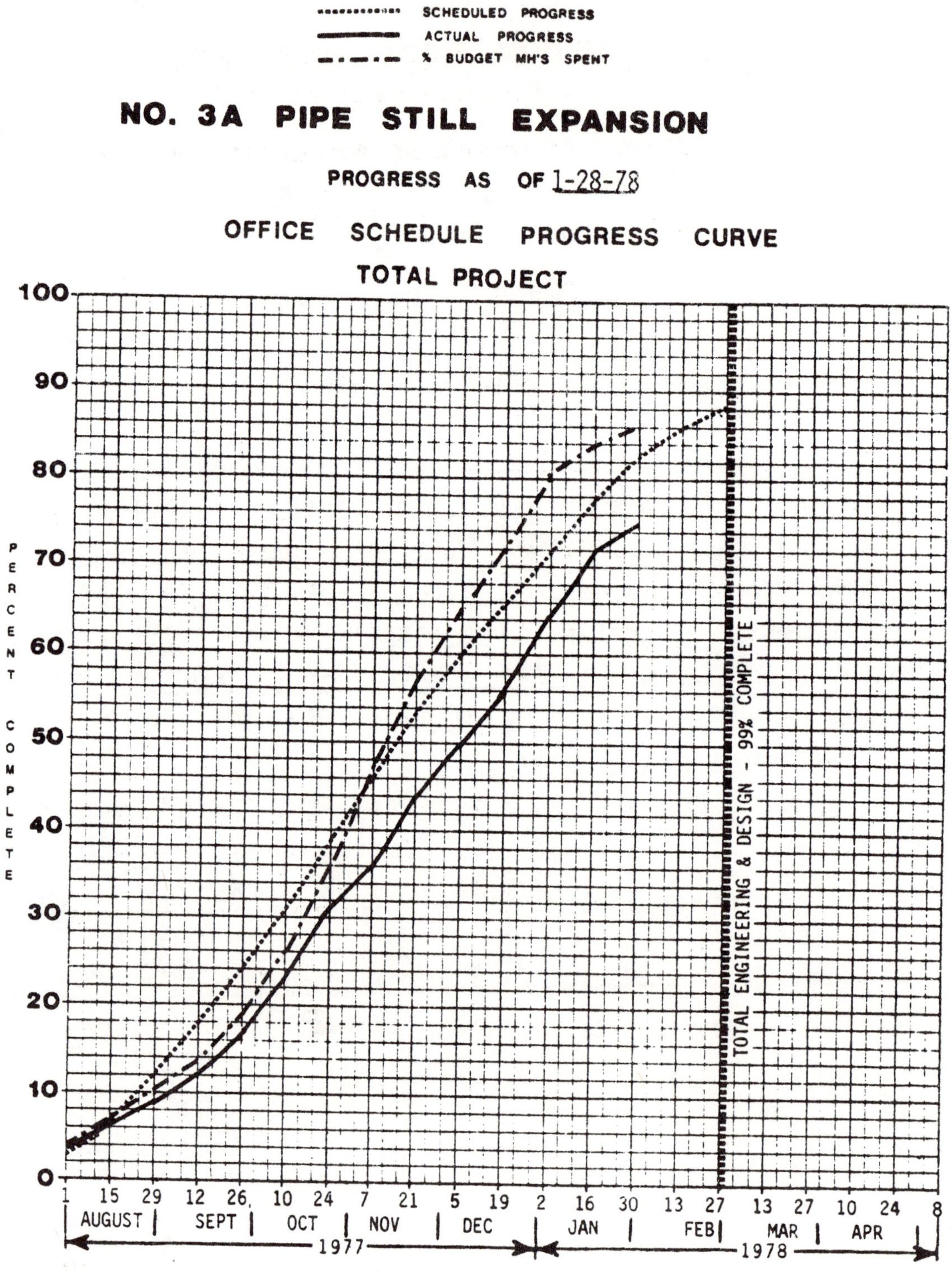

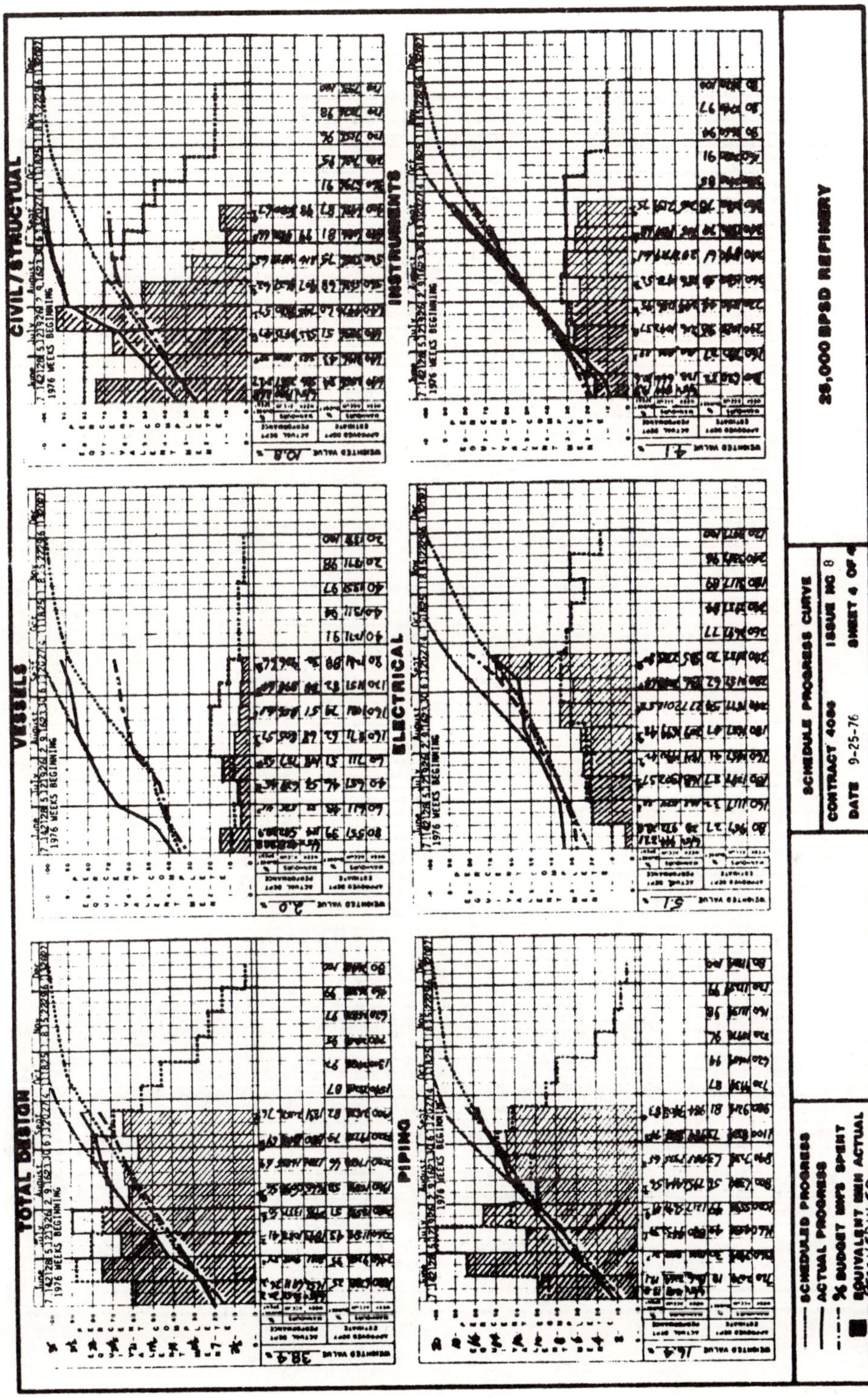

Fig. 6-20
Subsidiary curves

Fig. 6-21

Curve for Engineering

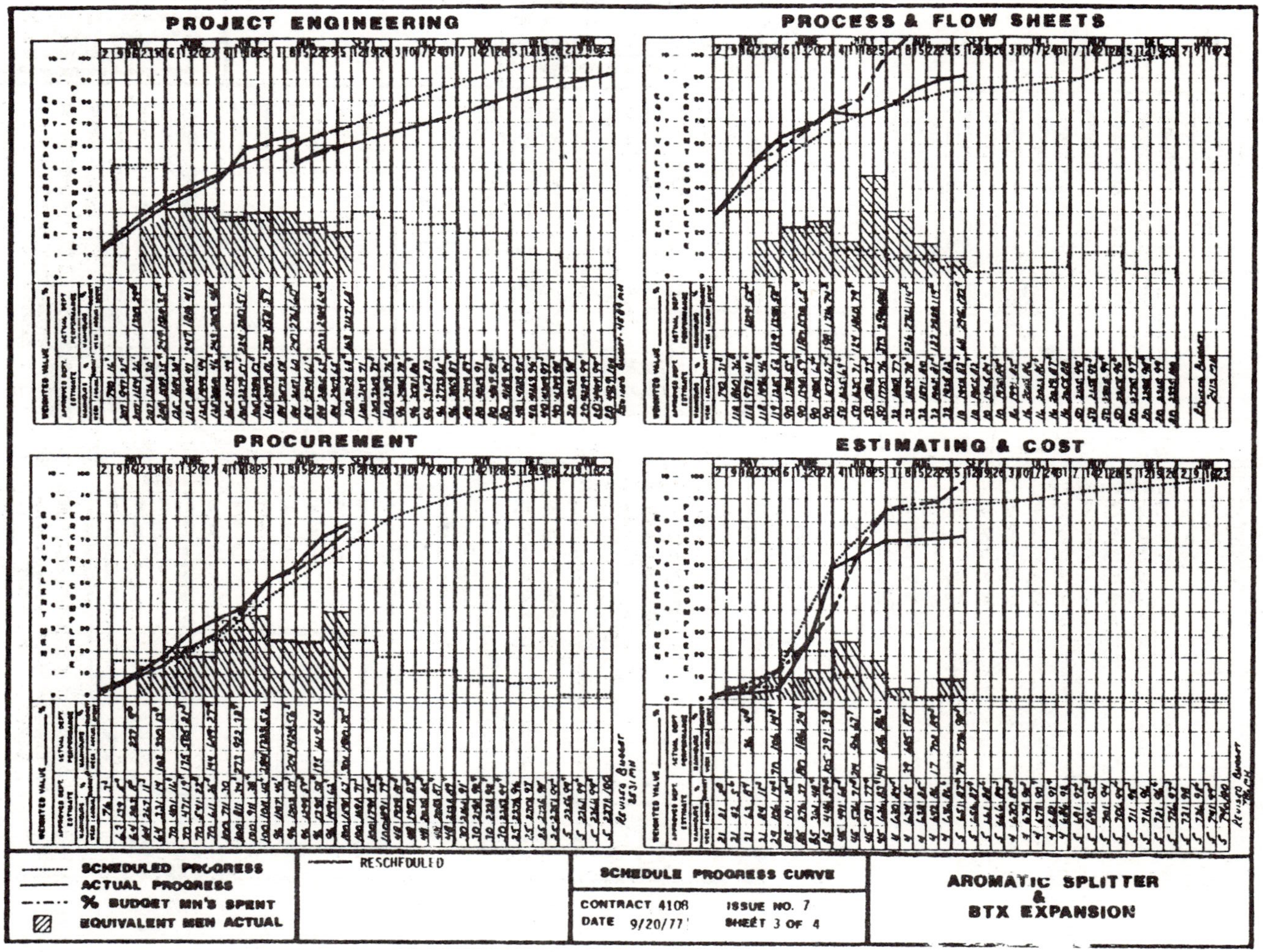

Fig. 6-22

Status report

STATUS REPORT
DATE: 08/16/77
EN7380301

CUSTOMER:
UNIT: BTX EXP + AROMATICS SPLTR
LOCATION:

REPORT NBR: 6
PER / END: 08/13/77

CONTRACT: 0410A
UNIT NBR: 0000
OPER NBR: 22

SCOPE CHANGE ITEM / DWG NBR	SCHEDULED DATES START	SCHEDULED DATES COMPLETE	ACTUAL DATES START	ACTUAL DATES COMPLETE	DESCRIPTION	EST HOURS	PCT COMP	ISSUE STATUS REV	ISSUE STATUS DATE	NOTE	NOTE CODE
A01	04/20/77	10/30/77	04/20/77		SUPERVISION, MISCELLANEOUS VENDOR DWG. LOGGING	280	85	CUR- AFC-			
A02	04/25/77	07/15/77	04/25/77	07/27/77	RAFFINATE WATER WASH COLUMN V-15	20	100	CUR- AFC-			
A03	04/25/77	07/15/77	04/25/77	07/27/77	EXTRACTOR WATER WASH COLUMN V-42	20	100	CUR- AFC-			
BTX-V-100	04/20/77	07/30/77	04/26/77	06/29/77	REWORK OF EXISTING BENZENE TOWER ITEM: V-1	40	100	CUR- 2 AFC-	06/29/77	THIS DWG IS NOT FOR CONSTRUCTION	
BTX-V-100V1	06/15/77	07/30/77			REWORK OF EXISTING BENZENE TOWER V-1 VENDOR CHECK (NOZZLES)	4		CUR- AFC-		LATE STARTING LATE COMPLETING ISSD FOR QUOTE	S02
BTX-V-100V2	06/15/77	07/30/77	07/18/77		REWORK OF EXISTING BENZENE TOWER V-1 VENDOR CHECK (TRAYS)	8	80	CUR- AFC-		LATE COMPLETING	
BTX-V-101	06/22/77	08/15/77	06/15/77	08/11/77	REWORK OF EXISTING TOLUENE COLUMN ITEM: V-3	20	100	CUR- 0 AFC-	06/29/77	THIS DWG IS NOT FOR CONSTRUCTION	
BTX-V-101VC	07/15/77	08/15/77	08/11/77	08/11/77	REWORK OF EXISTING TOLUENE COLUMN VENDOR CHECK (VESSEL) V-3	8	100	CUR- AFC-			
BTX-V-102	06/06/77	06/15/77	06/23/77	08/11/77	REWORK OF EXISTING EXTRACTOR ITEM: V-13	25	100	CUR- 0 AFC-	06/29/77	THIS DWG IS NOT FOR CONSTRUCTION	
BTX-V-102V1	07/01/77	08/15/77	08/11/77	08/11/77	REWORK OF EXIST. EXTRACTOR V-13 VENDOR CHECK (NOZZLES)	8	100	CUR- AFC-			
BTX-V-102V2	07/15/77	08/15/77	05/25/77		REWORK OF EXIST. EXTRACTOR V-13 VENDOR CHECK (TRAYS)	30	90	CUR- AFC-		DUE TO COMPLETE	
BTX-V-103	06/06/77	08/15/77	06/23/77	08/11/77	MISCELLANEOUS DETAILS OF EXIST. EXTRACTOR ITEM: V-13	15	100	CUR- 0 AFC-	06/29/77	THIS DWG IS NOT FOR CONSTRUCTION	

REPORT NBR: 6
PAGE NBR: 11

Fig. 6-23

Project status summary

PROJECT STATUS SUMMARY
DATE: 09/27/77
EN73R0302

CUSTOMER:
PROJECT: NO 3A PIPE STILL EXP - GOF
LOCATION:

REPORT NBR: 6
PER / END: 09/24/77

CONTRACT: 4119

OPERATION DESCRIPTION	OPER NBR	PCT COMP	SCHED PCT COMP	+AHEAD -BEHIND SCHED	PCT GAIN THIS PER	CURR HRS	TOTAL HRS TO DATE	FRCST HOURS TO GO	TOTAL HOURS FRCST	BUDGET HOURS	PCT BGT EXPND
PLOT PLANS	21	90	80	+10	0	0	14	116	130	100	14
VESSEL ENGINEERING	22	39	39		9	84	329	51	390	390	84
PIPING ENGINEERING	23	1	10	-9	0	6	27	335	362	350	8
PIPING DESIGN	24	10	10		5	180	360	1642	2002	2030	18
PIPING ISOMETRICS	25	0	0		0	0	0	1870	1870	2020	0
MATL CONTROL - PIPING	26	2	1	+1	0	0	16	398	414	410	4
CIVIL/STRUCT ENG	27	34	34		13	156	364	386	750	560	65
CIVIL/STRUCT DESIGN	28	6	10	-4	2	24	85	1270	1355	1050	8
ELECTRICAL ENGINEERING	31	21	21		2	18	107	499	606	410	26
ELECTRICAL DESIGN	32	6	6		4	54	85	615	700	710	12
MATL CONTROL - ELEC	33										
INSTRUMENT ENGINEERING	34	12	12		7	115	214	1135	1349	880	24
INSTRUMENT DRAFTING	35	0	2	-2	0	17	36	754	790	490	7
PLANT MODELS	37										
REQUISITION REPORT	--										
TOTAL ENGINEERING OPERATIONS:		10½	10½			654	1637	9081	10718	9400	17
ESTIMATING	10	0				0	0			0	0
COST CONTROL	11	27	27		4	12	78	357	435	250	31
SCHEDULING	12	32	32		7	32	145	125	270	280	52
PURCHASING	13	11	11		6	32	56	504	560	560	10
EXPEDITING	14	0	2	-2	0	21	28	462	490	490	6
INSPECTION	15	6	6		0	38	79	361	440	440	18
TRAFFIC	16										
VENDOR PRINT CONTROL	17	0				0	1	589	590	0	0
RECR + CL - REIMB	18	22	18	+4	8	5	55	1295	1350	2995	2
RECR + CL - NONREIMB	19					65	94	0	94	0	0
FLOW SHEETS	20	78	78		19	22.	195	40	235	130	150
FLD CK+ST-UP -NONREIMB	39										
SECTION CHIEFS	40	0				0	23	0	23	0	0
ENGINEERING MANAGEMENT	41					1	6	0	6	0	0
FLD CHECKOUT - REIMB	42	0									
PROJECT ENGINEERING	53	31	31		5	60	428	1088	1516	1460	29
PROJECT MANAGEMENT	54	22	21	+1	3	40	389	817	1206	2270	17
GENERAL PROCESS	56	72	72		13	22	433	32	465	190	241
OPERATING MANUALS	57										
FLD ST-UP - REIMB	58										
TOTAL PROJECT OPERATIONS:						370	2010	5730	7740	9045	22
TOTAL PROJECT:						1024	3647	14811	18458	18445	20

Fig. 6-24

LEGEND:
Planned.......
Actual ——

PROJECT 40

25,000 BPSD REFINERY
Progress as of 8/22/77

TOTAL CONSTRUCTION PROGRESS CURVE

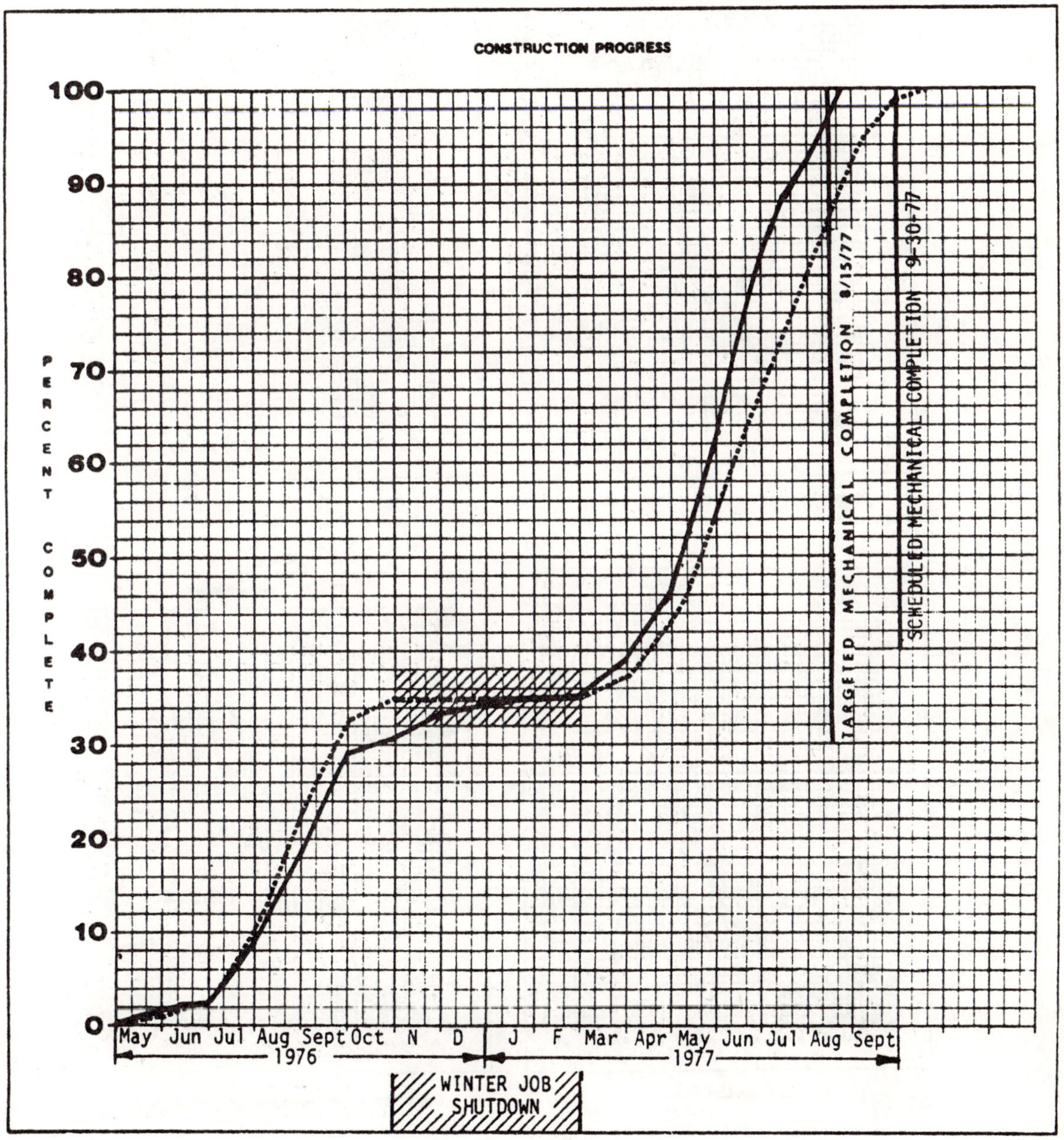

Fig. 6-25

LEGEND:
Planned ....
Rescheduled ~~~~~
Actual ——
% MHrs —·—

PROJECT 40
Rev. 3 5/10/77

25,000 BPSD REFINERY
Progress as of 7-30-77

PHASE II (1977) CONSTRUCTION PROGRESS CURVES (PG 1/2)

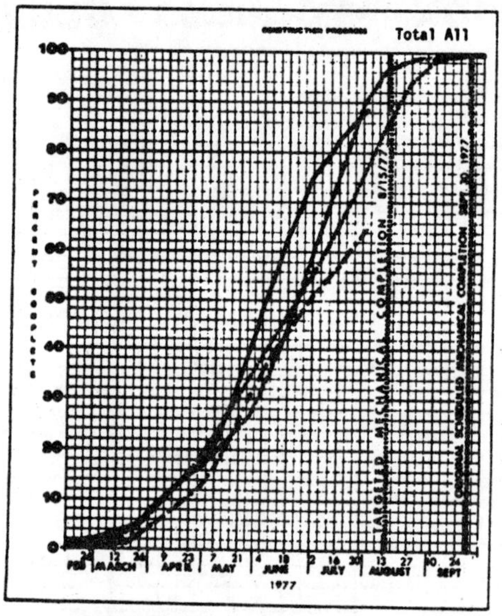

Total 1977 (Phase II) Construction

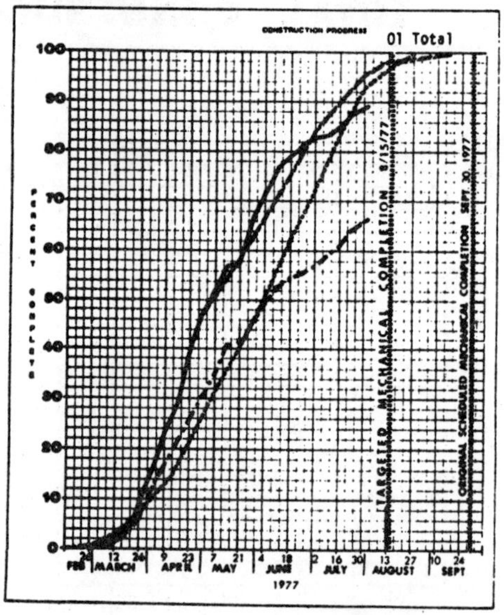

Howe-Baker Crude Unit Erection

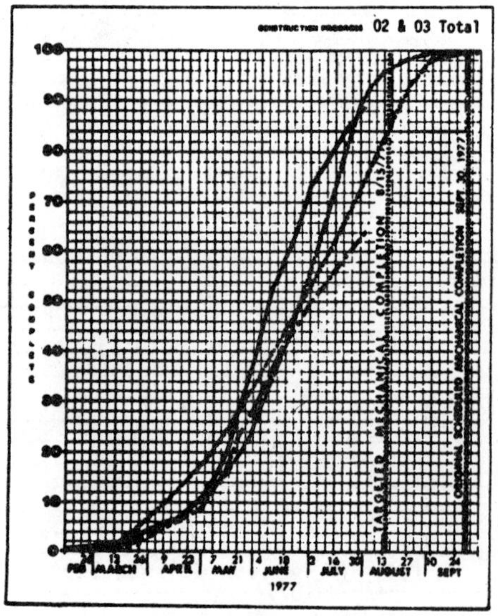

Offsites Erection

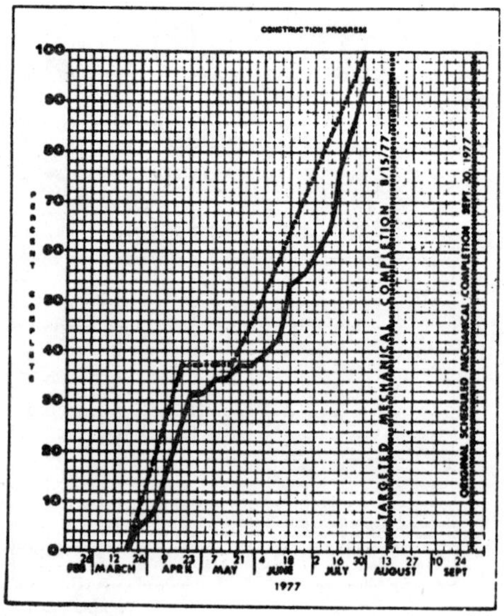

Pipeline & Metering Bldg Construction

Fig. 6-26

Planned.......
Actual ——
o/o MHrs .—..

PROJECT 4C

25,000 BPSD REFINERY

Progress as of 7-[illegible]-77

ELECTRICAL SUBCONTRACTOR

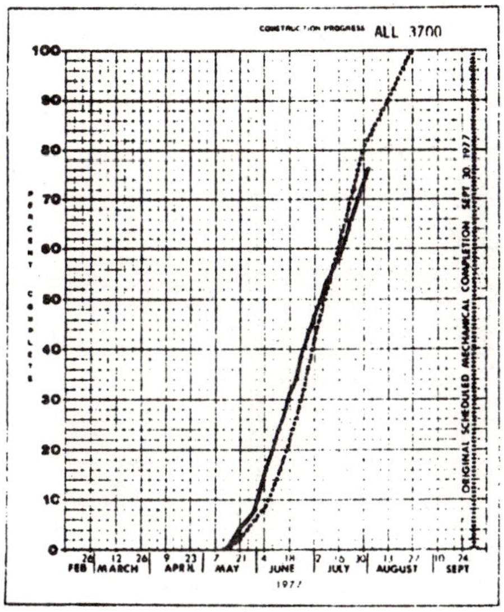

Total Electrical Work

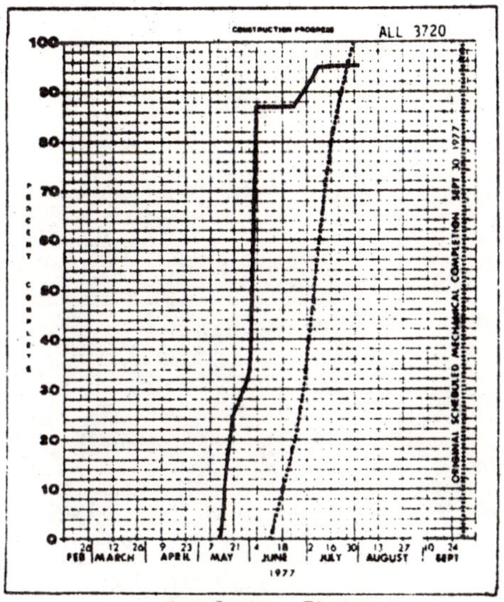

Set/Connect Electrical Equipment in Buildings

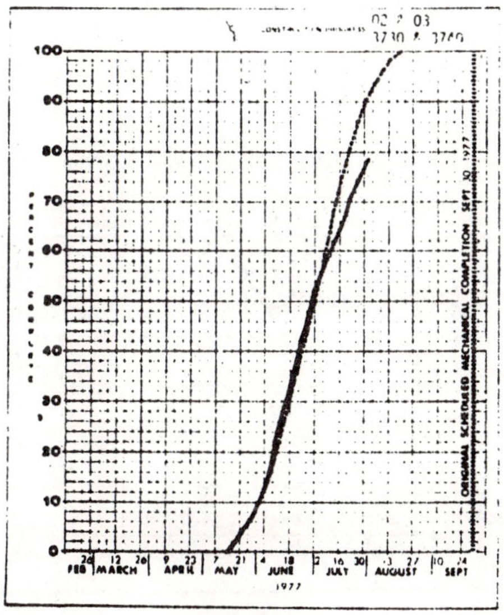

Offsites Work-ISBL & OSBL

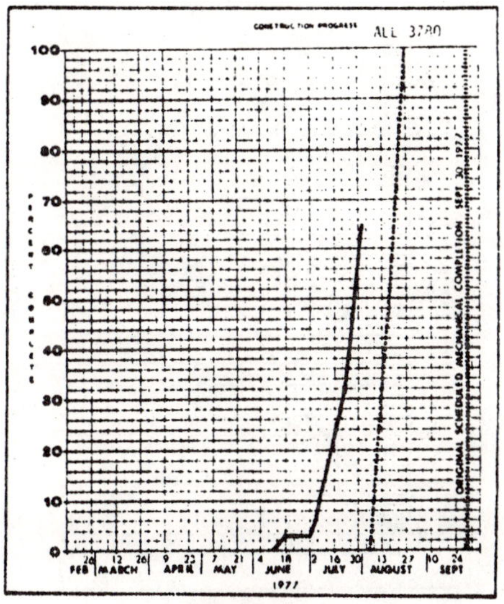

Energize & Checkout Refinery

FINAL ISSUANCE OF THIS PAGE

Fig. 6-27

PROJECT 40
REV 2 3/24/77

25,000 BPSD REFINERY

Progress as of 7-30-77

FIELD INDIRECT COSTS (PG 1/2)

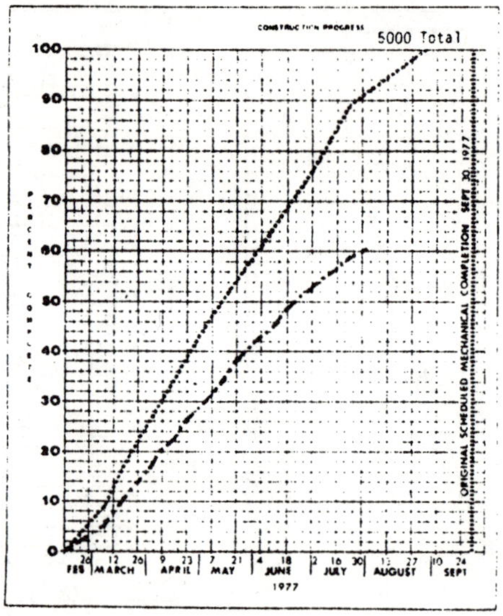

TOTAL FIELD INDIRECT COSTS

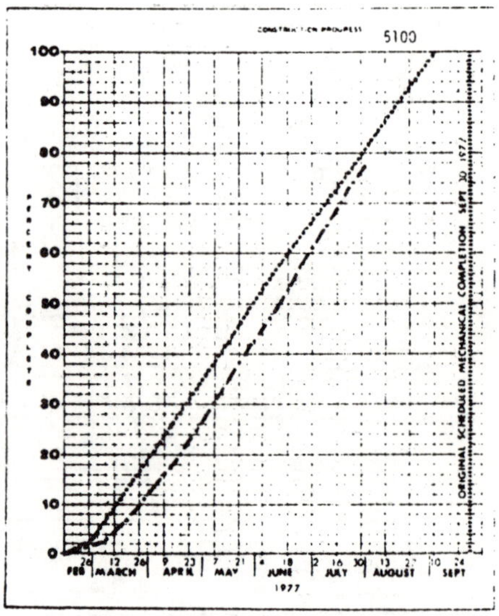

TOTAL FIELD OFFICE LABOR

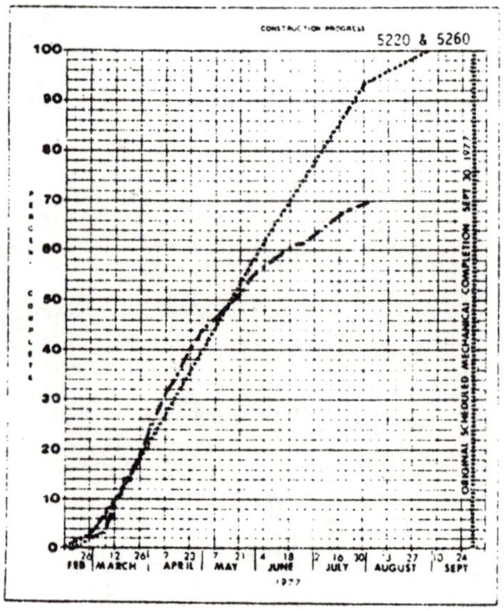

TOTAL WAREHOUSE & REMOTE HANDLING/STORAGE LABOR

Fig. 6-28
Progress report

RUN DATE: 09/28/77
PROGRAM NUMBER: 1
JOBUNIT: 040860003 OSBL OFFSITES

PIPE ERECTION PROGRESS

REPORT DATE: 2/12/7
PAGE NUMBER: 7
REPORT NUMBER: 50

LARGE BORE QUANTITIES

LINE	SH	SHOP FEET TL	SHOP FEET TD	SHOP FEET RM	SHOP FEET 51	FIELD FEET TL	FIELD FEET TD	FIELD FEET RM	FIELD FEET 51	BOLT-UPS TL	BOLT-UPS TD	BOLT-UPS RM	BOLT-UPS 51	WELDS TL	WELDS TD	WELDS RM	WELDS 51
0323	2	0	0	0	...	0	0	0	...	0	0	0	...	0	0	0	...
0323	3	0	0	0	...	0	0	0	...	0	0	0	...	0	0	0	...
0341	1	0	0	0	...	0	0	0	...	0	0	0	...	0	0	0	...
0341	2	0	0	0	...	0	0	0	...	0	0	0	...	0	0	0	...
0366	4	0	0	0	...	0	0	0	...	0	0	0	...	0	0	0	...
0366	5	0	0	0	...	0	0	0	...	0	0	0	...	0	0	0	...
0366	6	0	0	0	...	0	0	0	...	0	0	0	...	0	0	0	...

SMALL BORE QUANTITIES

LINE	SH	FEET TL	FEET TD	FEET RM	FEET 51	BOLT-UPS TL	BOLT-UPS TD	BOLT-UPS RM	BOLT-UPS 51
0323	2	20	20	0	...	0	0	0	...
0323	3	59	59	0	...	0	0	0	...
0341	1	59	59	0	...	0	0	0	...
0341	2	172	172	0	...	0	0	0	...
0366	4	294	294	0	...	0	0	0	...
0366	5	405	405	0	...	0	0	0	...
0366	6	304	344	-40	...	0	0	0	...

SMALL BORE QUANTITIES (WELDS) / MANHOURS

LINE	SH	WELDS TL	WELDS TD	WELDS RM	WELDS 51	TO DATE BUD	TO DATE ERN	TO DATE SPT	RPT 50 ERN	RPT 50 SPT
0323	2	3	3	0	...	8	8	10	0	05
0323	3	50	50	0	...	53	53	22	0	05
0341	1	6	6	0	...	22	22	27	0	05
0341	2	15	15	0	...	64	64	5	0	05
0366	4	18	18	0	...	104	104	58	0	0
0366	5	46	46	0	...	153	153	18	0	0
0366	6	35	13	22	...	115	119	43	0	0
		21	21	0	...	123	123	30	0	0
		12	12	0	...	65	65	22	0	0
		75	75	0	...	253	253	104	0	0
		33	33	0	...	159	159	33	0	0
		10	5	5	...	34	41	8	0	0
		17	3	14	...	17	3	12	0	0
		60	25	35	...	139	103	31	0	0
		3	3	0	...	26	26	23	0	05
		4	2	2	...	56	54	34	0	05
		2	2	0	...	91	91	71	0	05
		6	6	0	...	67	67	39	0	05
		7	7	0	...	21	21	24	0	05
		8	8	0	...	47	47	31	0	05
		18	18	0	...	72	72	31	0	05
		4	4	0	...	37	37	70	0	05
		8	8	0	...	117	117	181	0	05

LITWIN CORPORATION
FIELD REMAINING WORK SUMMARY

	QUANTITIES TOTAL	QUANTITIES TODATE	QUANTITIES REMAIN	QUANTITIES %	MANHOURS BUDGET	MANHOURS EARNED	MANHOURS REMAIN	MANHOURS %
LARGE SHOP HIGH	0	0	0	0.0	0.	0.	0.	0.0
LARGE SHOP LOW	11414	11403	11	99.9	5434.	5430.	4.	99.9
LARGE FIELD	2228	2176	52	97.7	966.	950.	17.	98.3
LARGE BOLT-UPS	611	597	14	97.7	1645.	1599.	46.	97.2
LARGE WELDS	429	415	14	96.7	1333.	1295.	38.	97.1
SMALL FEET	10029	9874	155	98.5	3203.	3154.	49.	98.5
SMALL BOLT-UPS	63	58	5	92.1	91.	87.	5.	95.0
SMALL WELDS	1650	1506	144	91.3	1031.	954.	78.	92.5
ALL LARGE BORE					9378.	9273.	105.	98.9
ALL SMALL BORE					4326.	4194.	131.	97.0
TOTAL					13704.	13468.	236.	98.3

TESTING TLF	TESTING SLF	TESTING %
23671	17095	72.2

Fig. 6-29

Construction progress report, direct labor

F 08/29/77
1W0401
FMT:
TRACT: 04086.0003 OSBL OFFSITES

CONSTRUCTION PROGRESS REPORT
DIRECT LABOR

PAGE: 1
PERIOD ENDING: 08/28/77

				BUDGET			ACTUAL						FORECAST		
T	CODE	DESCRIPTION		UNITS	MAN-HRS / UNIT	MAN HOURS	TOTAL UNITS	UNITS TODATE	PHY PCT COMP	PCT BUD EXPD	MAN-HRS TODATE	MAN-HRS / UNIT	MAN-HRS / UNIT	TOTAL MAN-HRS	M/H PCT COMP
31		DRUMS HAUL + SET	TN	18	6.00	108	18	18.00	100.0	25.9	28.0	1.56	1.56	28F	100.0
32		DRUMS INSPECT + CLEAN	EA	1	12.00	12	1	1.00	100.0				1.00	1F	
	2130	TOTAL DRUMS				120*			100.0	23.3	28.0			29	96.5
[illegible]		PROCESS SKIDS	EA	3	63.00	189	3	3.00	100.0	125.1	236.5	78.83	79.00	237F	99.7
3		PIPE SKIDS	EA	46	40.00	1840	46	46.00	100.0	66.8	1230.0	26.74	26.74	1230F	100.0
	2380	TOTAL MISC EQUIP				2029*			99.9	72.2	1466.5			1467	99.9
53		LADDERS + PLATFORMS	TN	7	70.43	493	7	7.00	100.0	69.3	342.0	48.86	48.86	342I	100.0
54		PIPE RACKS + SLEEPERS	TN	43	20.00	860	43	43.00	100.0	147.8	1271.5	29.57	29.58	1272F	99.9
59		MISC I. SUPPORTS	TN	1	40.00	40	1	1.00	100.0	100.0	40.0	40.00	40.00	40F	100.0
	2460	TOTAL STRUCTURAL STEEL				1393*			99.9	118.7	1653.5			1654	99.9
21		SMALL BORE ERECTION	LF	9846	.71	6991	0029	10029.00	100.0	51.1	3574.5	.36	.36	3575F	99.9
22		SMALL BORE FLD FAB WLDS	EA	2800	.50	1400	3252	3252.00	100.0	147.3	2063.0	.63	.63	2063F	100.0
23		LARGE BORE ERECTION	LF	13613	1.00	13596	3612	13612.00	100.0	68.9	9371.0	.69	.69	9371L	100.0
	2520	TOTAL CS PIPE				21987*			99.9	68.2	15008.5			15009	100.0
75		BRACKETS + HANGERS	EA	752	5.00	3760	927	927.00	100.0	35.0	1318.0	1.42	1.42	1318I	100.0
77		STEAM/HOT OIL TRACING	LF	10148	.62	6263	0148	10148.00	100.0	76.4	4788.5	.47	.47	4788I	100.0
	2570	TOTAL MISC PIPING				10023*			99.9	60.9	6106.5			6106	100.0
11		FIELD INSTRUMENTS	EA	52	8.75	455	54	54.00	100.0	77.1	351.0	6.50	6.50	351F	100.0
12		TRANSMITTERS	EA	14	33.00	462	6	6.00	100.0	11.9	55.0	9.17	9.17	55F	100.0
	2610	TOTAL FIELD INSTRUMENTS				917*			99.9	44.2	406.0			406	100.0
542		INSTRUMENT PANELS	EA	1	55.00	55	1	1.00	100.0				1.00	1F	
	2640	TOTAL BOARD INSTR + PANEL				55*			100.0					1	
581		LOOP CHECK	*	100	5.42	542	100	100.00	100.0	11.8	64.0	.64	.64	64F	100.0
	2680	TOTAL INSTR LOOP CHECK +				542*			100.0	11.8	64.0			64	100.0
721		POWER EQUIPMENT	TN	5	20.00	100	5	5.00	100.0	14.0	14.0	2.80	2.80	14F	100.0
	2720	TOTAL POWER EQUIPMENT				100*			100.0	14.0	14.0			14	100.0
932		TESTING A/G PIPE	*	100	33.00	3300	100	100.00	100.0	59.4	1962.5	19.63	19.63	1963F	99.9
	2930	TOTAL TESTING + CHEM. CLEA				3300*			100.0	59.4	1962.5			1963	99.9
	2000	TOTAL LITWIN DIRECT LABOR				42534			100.0	68.6	29215.0			31351	93.1

Fig. 6-30

FL61W0401
CLIENT:
CONTRACT: 04086.0003 OSBI OFFSITES

CONSTRUCTION PROGRESS REPORT
SUB-CONTRACT DIRECT LABOR

PAGE: 3
PERIOD ENDING: 07/17/77

COST CODE	DESCRIPTION		BUDGET UNITS	BUDGET MAN-HRS / UNIT	BUDGET MAN HOURS	ACTUAL TOTAL UNITS	ACTUAL UNITS TODATE	ACTUAL PHY PCT COMP	ACTUAL PCT BUD EXPD	ACTUAL MAN-HRS TODATE	ACTUAL MAN-HRS / UNIT	FORECAST MAN-HRS / UNIT	FORECAST TOTAL MAN-HRS	FORECAST M/H PCT COMP
3415	SHAPING	SY	18620	.02	282	18620						.02	282F	
3417	DIKE LINERS	SF	160000		400	160000							400F	
3419	DIKE WORK	CY	7410	.07	550	7410	1630.00	22.0	21.8	120.0	.07	.07	550F	21.8
3410	TOTAL EARTHWORK				1232			12.8	9.7	120.0			1232	9.7
3580	CULVERTS	LF	132	.50	66	132						.50	66E	
3580	TOTAL UG PIPE				66								66	
3714	GROUNDING	LF	1036	.08	83	1036	32.00	3.0	33.7	28.0	.88	.08	83E	33.7
3710	TOTAL UG ELECTRICAL				83			3.0	33.7	28.0			83	33.7
3721	POWER EQUIPMENT	EA	18	4.44	80	18	18.00	100.0	100.0	80.0	4.44	4.44	80F	100.0
3720	TOTAL POWER EQUIP				80			100.0	100.0	80.0			80	100.0
3731	ABOVEGROUND CONDUIT	LF	20000	.13	2600	15300	12240.00	80.0	80.0	2080.0	.17	.17	2600F	80.0
3732	CABLE + WIRE	LF	155000	.01	1550	155000	32500.00	20.9	20.9	325.0	.01	.01	1550F	20.9
3733	TRAYS	LF	1200	.20	240	1200	1080.00	90.0	91.6	220.0	.20	.20	240F	91.6
3734	INSTRUMENT CABLE	LF	15000	.01	150	15000	5250.00	35.0	33.3	50.0	.01	.01	150F	33.3
3735	ELECTRICAL TRACING	LF	7920	.10	792	7920	6495.00	82.0	82.0	650.0	.10	.10	792F	82.0
3730	TOTAL POWER DISTRIB				5332			60.0	62.3	3325.0			5332	62.3
3741	LIGHTING EQUIPMENT	EA	127	3.00	381	127	51.00	40.1	40.6	155.0	3.04	3.00	381F	40.6
3740	TOTAL LIGHTING EQUIP				381			40.1	40.6	155.0			381	40.6
3781	ELECTRICAL CHECKOUT	$	100	9.42	942	100	10.00	10.0	10.0	95.0	9.50	9.42	942F	10.0
3780	TOTAL ELEC TSTG + CAL				942			10.0	10.0	95.0			942	10.0
3811	PIPE INSULATION	LF	23674	.30	7200	23674	7291.00	30.8	30.7	2212.0	.30	.30	7200F	30.7
3812	EQUIPMENT INSULATION	SF	2030	.05	100	2030						.05	100F	
3815	TANK INSULATION	SF	83140	.08	6400	83140	15060.00	18.1	18.1	1160.0	.08	.08	6400F	18.1
3810	TOTAL INSULATION				13700			24.4	24.6	3372.0			13700	24.6
3832	PIPE PAINTING	LF	11874	.05	577	11874	215.00	1.8	1.7	10.0	.05	.05	577F	1.7
3833	EQUIPMENT PAINTING	SF	36400	.04	1470	36400	26040.00	71.5	71.4	1050.0	.04	.04	1470F	71.4
3834	STEEL PAINTING	SF	15300	.02	230	15300	4160.00	27.1	28.2	65.0	.02	.02	230E	28.2
3830	TOTAL PAINTING				2277			48.1	49.4	1125.0			2277	49.4
3000	TOTAL SUB-CONTR DIRECT LA				24093			33.3	34.4	8300.0			24093	34.4
2000/3000	TOTAL DIRECT AND SUB-CONT				66627			72.2	51.7	34496.5			57878	59.6

Fig. 6-31

Construction progress report, labor summary

FL61W0403
CLIENT:
CONTRACT: 04086 ALL UNITS

CONSTRUCTION PROGRESS REPORT
LABOR SUMMARY

PAGE: 1
PERIOD ENDING: 06/12/77

			ACTUAL			FORECAST			PERCENT COMPLETE				
COST CODE	DESCRIPTION	BUDGET MAN HRS	THIS PER MAN HRS	TO DATE MAN HRS	TO GO MAN HRS	TOTAL MAN HRS	CHANGE MAN HRS	PHY PCT CUR PER	MH PCT CUR PER	PHY PCT TODATE	BD PCT EXPEND	MH PCT TODATE
2110	TOWERS	506		215.0		215				99.9	42.4	100.0
2130	DRUMS	618		307.0	50	357				97.3	49.6	85.9
2100	TOTAL VESSELS LABOR	1124*		522.0	* 50	572*				98.5	46.4	91.2
2210	S + T EXCHANGERS	699		199.0		199				99.9	28.4	100.0
2230	AIR COOLED EXCHANGERS	1406	61.0	667.5	488	1156		2.9	5.2	65.9	47.4	57.7
2240	FIRED HEATERS	4658	81.0	2153.5	104	2257		.1	3.5	96.8	46.2	95.3
2250	BOILERS	347	5.0	797.5	102	900		4.4	.5	95.5	229.8	88.6
2200	TOTAL HEAT TRANSFER EQ LA	7110*	147.0	3817.5	* 695	4512*		.8	3.2	90.9	53.6	84.5
2310	PUMPS + DRIVERS	921	28.5	428.5	403	832			3.4	69.3	46.5	51.5
2330	ELECTRICAL GEN + DR	100		44.0		44				100.0	44.0	100.0
2370	MIXERS + AGITATORS	50	23.0	35.0		35	+6	25.0	65.7	100.0	70.0	100.0
2380	MISC EQUIP	5247	194.0	4583.0	1116	5699		1.3	3.4	93.9	87.3	80.4
2300	TOTAL MECHANICAL EQUIP LA	6318*	245.5	5090.5	* 1519	6610*	+6	1.3	3.7	91.0	80.5	77.0
2430	UNDERGROUND CONCRETE	1561	415.0	2596.0	311	2907	+379	9.5	14.2	97.3	166.3	89.3
2460	STRUCTURAL STEEL	4520	120.0	3505.0	401	3906	-22	7.2	3.0	95.2	77.5	89.7
2470	BUILDINGS	38		16.0		16				100.0	42.1	100.0
2400	TOTAL CIVIL LABOR	6119*	535.0	6117.0	* 712	6829*	+357	7.8	7.8	95.8	99.9	89.5
2520	CS PIPE	45964	2357.0	26526.5	7080	33607	-6104	6.9	7.0	83.1	57.7	78.9
2570	MISC PIPING	23525	1626.5	5436.5	16631	22068		14.3	7.3	40.1	23.1	24.6
2500	TOTAL PIPING LABOR	69489*	3983.5	31963.0	* 23712	55675*	-6104	9.5	7.1	68.0	45.9	57.4
2610	FIELD INSTRUMENTS	1749	91.0	667.0	359	1026	-555	33.8	8.8	75.5	38.1	65.0
2640	BOARD INSTR + PANELS	220		15.0	205	220				5.0	6.8	6.8
2650	INSTR PIPING + TUBING	3457	304.5	1597.5	1528	3126	-461	24.7	9.7	61.1	46.2	51.1
2670	INSTR WEATHER PROTECTION	676		51.5	394	446				32.3	7.6	11.5
2680	INSTR LOOP CHECK + CAL	1626		33.0	1593	1626		3.3		3.3	2.0	2.0
2600	TOTAL INSTRUMENTS LABOR	7728*	395.5	2364.0	* 4080	6444*	-1016	19.3	6.1	48.8	30.5	36.6
2720	POWER EQUIPMENT	300		48.0		48				99.9	16.0	100.0
2700	TOTAL ELECTRICAL LABOR	300*		48.0	*	48*				99.9	16.0	100.0
2930	TESTING + CHEM CLEANING	3300	209.0	666.0	2634	3300		9.0	6.3	18.6	20.1	20.1
2940	CATALYSTS, CHEM + PCKG	100			100	100						
2950	MAINTENANCE, ETC EQUIP	331			331	331						
2960	FIRE PROTECTION EQUIP	376		67.0		67	-33	10.0		100.0	17.8	100.0
2980	DESCRIPTION NOT FOUND	100			100	100						
2990	PRECOMMISSIONING	3			3002	3002						
2900	TOTAL MISCELLANEOUS LABOR	4210*	209.0	733.0	* 6167	6900*	-33	7.9	3.0	23.4	17.4	10.6
2000	TOTAL LITWIN DIRECT LABOR	102398	5515.5	50655.0	36935	87590	-6790	12.3	6.2	69.6	49.4	57.8
3410	EARTHWORK	1232			1232	1232						
3440	RDS, PVG, SLBS	634			634	634						
3470	BUILDINGS	250		250.0		250				100.0	100.0	100.0
3400	TOTAL CIVIL SC	2116*		250.0	* 1866	2116*				14.0	11.8	11.8

To be effective, progress reporting on construction requires considerable detail and needs a field cost system that keeps up with all the detail. Fig. 6-28 is an example of a piping progress report. It shows each line number, large bore and small bore quantities, shop feet, field feet, bolt-ups, and welds, and the percentage of manhours budgeted, earned, and spent.

Summary reports for concrete, piping, electrical etc. are collected into an overall construction progress report that shows the various categories of work like equipment, piping, and instrumentation. Fig. 6-29 is such a report for direct labor. It shows the budget and actual manhours thus permitting a comparison between the two. It also permits forecasting for manhours for the total job. That procedure is monitoring the cost along the job and is cost control. Fig. 6-30 is a similar printout for sub-contracted direct labor. It is best to have subtotals for each contractor. Fig. 6-31 is a summary report for all the labor costs on the job.

The whole purpose of reporting is to avoid surprises and permit taking corrective action. Good field cost reporting is mandatory on a weekly basis, and the weekly report should be on the construction manager's desk by noon on Monday.

PERT, CPM, AND PRECEDENCE COMPUTERIZED SCHEDULING. In the early 1950s PERT, Programmed Evaluation and Review Technique was developed as a method for expediting completion of the U. S. Navy's Polaris missile program which has since been modified to include cost. CPM, Critical Path Method, was developed about the same time. Both use arrow diagrams for construction projects. In Fig. 6-32 the diagram CPM #1 illustrates the basic form of a network. Events (or nodes) are shown by circles

Fig. 6-32

Critical Path Method

CPM -1

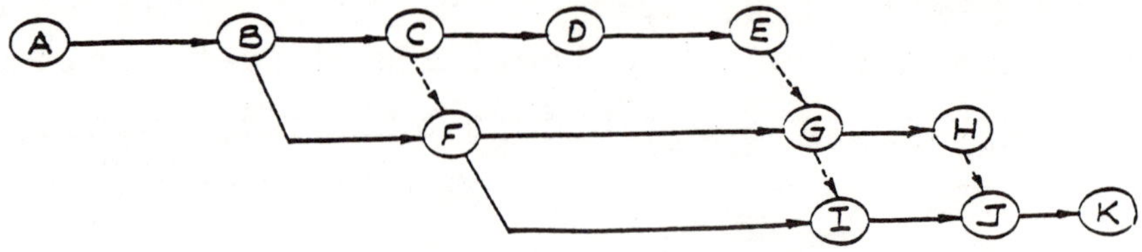

BASIC I,J LOGIC NETWORK

ACTIVITIES = AB, BC, CD, etc

RESTRAINTS = CF, EG, GI and HJ

CPM -2

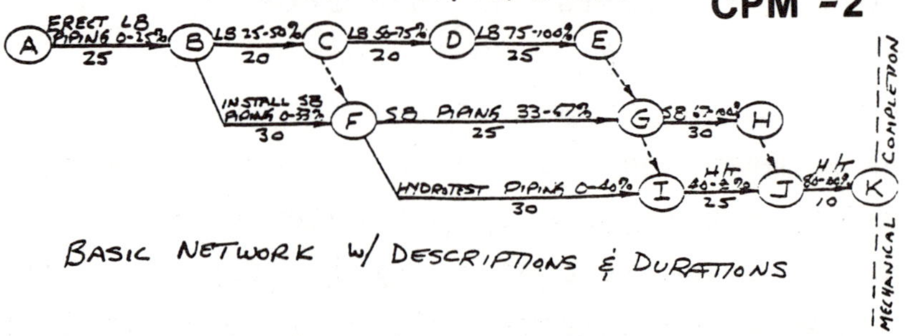

BASIC NETWORK W/ DESCRIPTIONS & DURATIONS

CPM -3

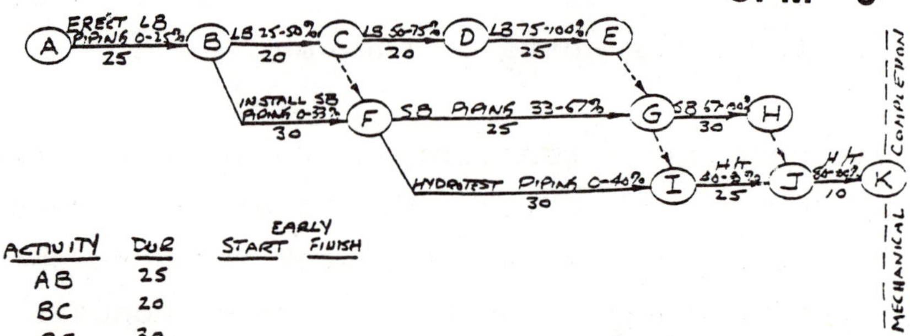

ACTIVITY	DUR	EARLY START	EARLY FINISH
AB	25		
BC	20		
BF	30		
CD	20		
CF	0		
DE	25		
EG	0		
FG	25		
FI	30		
GH	30		
GI	0		
IJ	25		
HJ	0		
KJ	10		

START OF BASIC NETWORK LISTING

Fig. 6-33

Critical Path Method

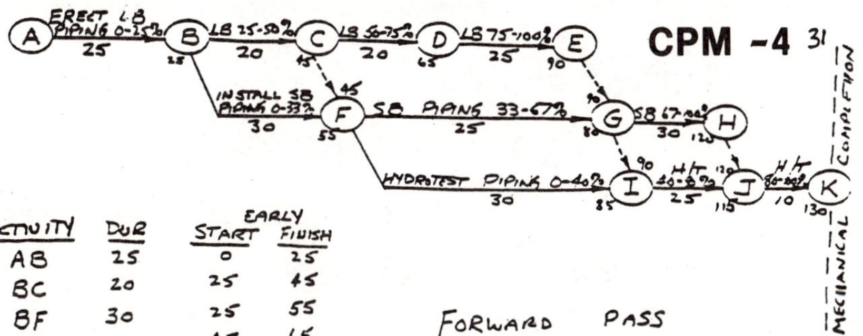

ACTIVITY	DUR	EARLY START	EARLY FINISH
AB	25	0	25
BC	20	25	45
BF	30	25	55
CD	20	45	65
CF	0	45	45
DE	25	65	90
EG	0	90	90
FG	25	55	80
FI	30	55	85
GH	30	90	120
GI	0	90	90
IJ	25	90	115
HJ	0	120	120
JK	10	120	130

FORWARD PASS CALCULATION OF EARLY START/FINISH

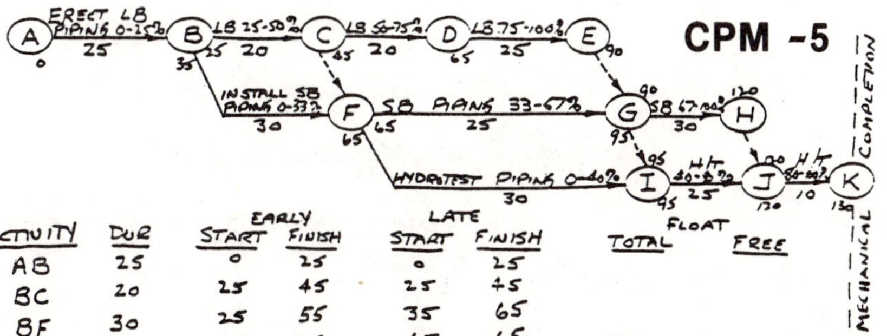

ACTIVITY	DUR	EARLY START	EARLY FINISH	LATE START	LATE FINISH	FLOAT TOTAL	FLOAT FREE
AB	25	0	25	0	25		
BC	20	25	45	25	45		
BF	30	25	55	35	65		
CD	20	45	65	45	65		
CF	0	45	45	55	• 55		
DE	25	65	90	65	90		
EG	0	90	90	90	90		
FG	25	55	80	65	90		
FI	30	55	85	65	95		
GH	30	90	120	90	120		
GI	0	90	90	95	95		
IJ	25	90	115	95	120		
HJ	0	120	120	120	120		
KJ	10	120	130	120	130		

BACKWARD PASS CALCULATION OF LATE START/FINISH

Fig. 6-34
Critical Path Method

32

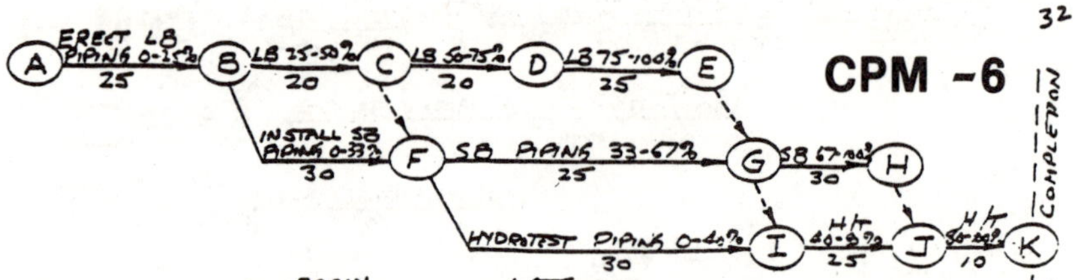

ACTIVITY	DUR	EARLY START	EARLY FINISH	LATE START	LATE FINISH	FLOAT TOTAL	FLOAT FREE
AB	25	0	25	0	25	0	
BC	20	25	45	25	45	0	
BF	30	25	55	35	65	10	
CD	20	45	65	45	65	0	
CF	0	45	45	55	55	10	
DE	25	65	90	65	90	0	
EG	0	90	90	90	90	0	
FG	25	55	80	65	90	10	
FI	30	55	85	65	95	10	
GH	30	90	120	90	120	0	
GI	0	90	90	95	95	5	
IJ	25	90	115	95	120	5	
HJ	0	120	120	120	120	0	
KJ	10	120	130	120	130	0	

CALCULATION OF TOTAL FLOAT

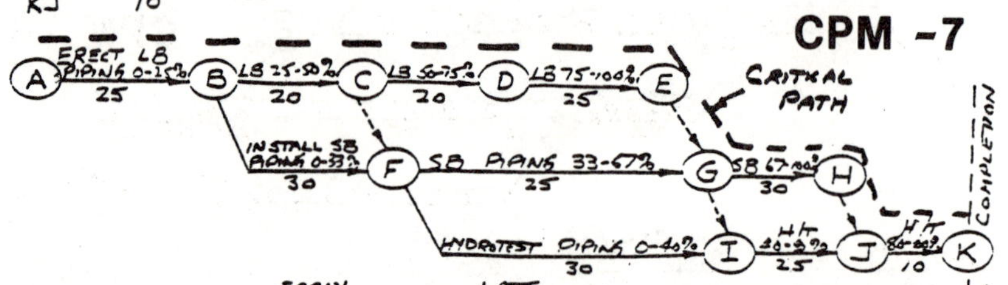

ACTIVITY	DUR	EARLY START	EARLY FINISH	LATE START	LATE FINISH	FLOAT TOTAL	FLOAT FREE
AB	25	0	25	0	25	0 -	
BC	20	25	45	25	45	0 -	
BF	30	25	55	35	65	10	
CD	20	45	65	45	65	0 -	
CF	0	45	45	55	55	10	
DE	25	65	90	65	90	0 -	
EG	0	90	90	90	90	0 -	
FG	25	55	80	65	90	10	
FI	30	55	85	65	95	10	
GH	30	90	120	90	120	0 -	
GI	0	90	90	95	95	5	
IJ	25	90	115	95	120	5	
HJ	0	120	120	120	120	0 -	
KJ	[illegible]	120	130	120	130	0 -	

CRITICAL PATH, i.e. "0" TOTAL FLOAT

Fig. 6-35

Critical Path Method

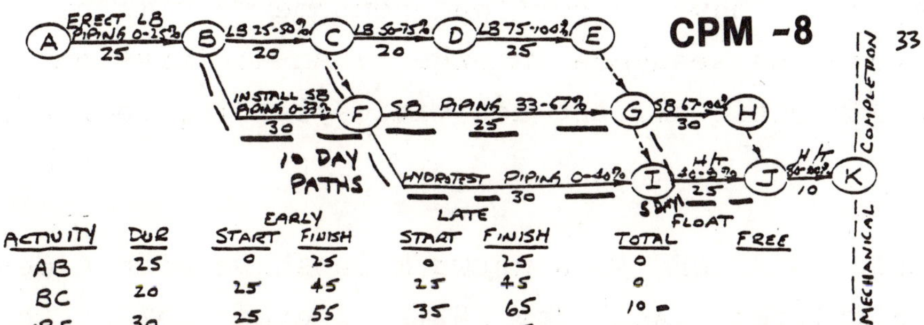

ACTIVITY	DUR	EARLY START	EARLY FINISH	LATE START	LATE FINISH	FLOAT TOTAL	FLOAT FREE
AB	25	0	25	0	25	0	
BC	20	25	45	25	45	0	
BF	30	25	55	35	65	10 -	
CD	20	45	65	45	65	0	
CF	0	45	45	55	55	10 -	
DE	25	65	90	65	90	0	
EG	0	90	90	90	90	0	
FG	25	55	80	65	90	10 -	
FI	30	55	85	65	95	10 -	
GH	30	90	120	90	120	0	
GI	0	90	90	95	95	5	
IJ	25	90	115	95	120	5	
HJ	0	120	120	120	120	0	
KJ	10	120	130	120	130	0	

10 DAY PATHS (SUB-CRITICAL)

5 DAY

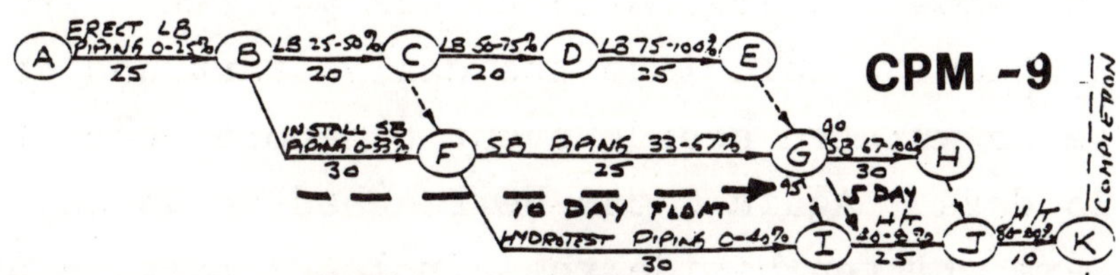

ACTIVITY	DUR	EARLY START	EARLY FINISH	LATE START	LATE FINISH	FLOAT TOTAL	FLOAT FREE
AB	25	0	25	0	25	0	
BC	20	25	45	25	45	0	
BF	30	25	55	35	65	10	
CD	20	45	65	45	65	0	
CF	0	45	45	55	55	10	
DE	25	65	90	65	90	0	
EG	0	90	90	90	90	0	
FG	25	55	80	65	90	10	
FI	30	55	85	65	95	10	
GH	30	90	120	90	120	0	
GI	0	90	90	95	95	5	
IJ	25	90	115	95	120	5	
HJ	0	120	120	120	120	0	
KJ	10	120	130	120	130	0	

like A start, intermediate events A, B, ... , and finish K. The arrows between events are called activities. The dotted arrows between C-F, E-G, and E-I are restraints, also called dummy activities, which have no cost and consume no time but are necessary to keep the sequence in order. An activity out of an event cannot begin until all the activities leading to that event are completed.

In Fig. 6-32 CPM #2 a network is shown with a description of the activities. Activity AB is to erect 25% of the large bore piping in 25 days. LB is large bore, SB is small bore, and HT is hydraulic testing. Dummy activities are shown with no duration. There is a milestone, namely mechanical completion at K.

In CPM #3 the duration for each activity is listed. The next step is shown in Fig. 6-33 as CPM #4. It is the forward pass calculation to calculate the early start and early finish dates. The 25 written beside event B denotes that B can be reached in 25 days. Again C can be reached in 45 days. At F both 45 and 55 are written beside the circle. Arriving by activity CF is 45 days and arriving by activity BF is 55 days. The larger number 55 prevails and no activities out of F can start before 55 days. When calculating the forward pass diagram always use the latest date at an event in moving forward. The total time to reach K, the finish, is 130 days. The early start and early finish for each activity are listed in the diagram.

The next step is to prepare the backward pass calculation to arrive at the latest start and latest finish dates as shown in CPM #5. The calculation is made working backwards from K, the finish. In the backward pass calculation when there are two or more numbers leading out of an event, the lowest number prevails. Thus at G in CPM #5 the start of GH shows 90 and the start of GI shows 95. The latest end for activities leading to G is 90 days if K is to be reached in 130 days. Working backwards it is seen that activity FG must start at 90 - 25 = 65 days. CPM #5 lists all the early and late dates for all activities.

In CPM #6 in Fig. 6-34 the early and late starts are compared. Consider the situation for activity BF. All activities leading to F can be finished in 55 days the earliest finish. All activities leading out of F can start as late as 65 days. Hence activity BF could take 10 days longer than the 25 allotted without delaying the finish of 130 days at K. There is 10 days of *total float* along the path BF. CPM #7 lists the total float along the activities. Activities having zero total float must be finished in the alloted time and are on the critical path for the network. In Fig. 6-35 CPM #8 shows the critical path ABCDEGHJK which takes 130 days to go from A to K. There are other paths but they are sub-critical and have some total float. They are a 5 day total float on path GIJ and a 10 day total float on each of the paths BF, CF, FG, and FI.

Total float is the time an activity can be delayed without affecting the completion of a project. Another term used is *free float* representing the time an activity can be delayed without affecting the earliest start of *any* other activity. This is shown in CPM #9. The activity FG is not on the critical path. It has a total float of 10 days. It also has free float. It must finish at 95 days and can start as late as 65 days. The difference is 30 days but the duration for FG is only 25 days. Path FG can be prolonged by 5 days, the free float, without influencing the earliest start of any other event.

In practice CPM is programmed on a computer and actual dates instead of elapsed days are printed out thus allowing for weekends, holidays and the like.

Schedules may be milestone or target schedules. A milestone is some key event like finish engineering or mechanical completion of the job. A milestone schedule picks out a small number of key events, say 10 to 20 out of several thousand, for reporting to management. A target schedule sets a schedule at the beginning of a project and reports actual dates against the target dates. The target schedule shows the original schedule against the schedule actually occurring.

A list of some terms used in CPM with their definitions follows:

activity	A task to be accomplished like pouring foundations. It is an operation that consumes time and other resources
arrow	A directed line between two events representing an actual or dummy activity
backward pass	A procedure working from finish to start establishing the latest finish and start times for the activities. See p. 6-37
bar chart	A chart on which activities are represented by bars showing start and finish
critical activity	An activity on the critical path
critical path	The longest duration chain of activities from start to finish. There is no float on the critical path

dangle	An activity for which the head or tail does not connect to an event. Not permitted
decision network	A network containing alternative paths
dummy activity	An activity essential for sequencing but having no duration
duration	The time required to complete an activity
event	A state in the diagram where certain activities must end before other activities can begin
E.F.T.	Earliest finish time of an activity
E.S.T.	Earliest start time of an activity
float	A time available for an activity in addition to its duration. May be negative
forward pass	A procedure working form start to finish establishing the earliest start and finish times for the activities. See p. 6-37
imposed date	A time set by circumstances outside the network
independent float	The time by which an activity may be delayed or extended without affecting preceding or suceeding activities
interface	An activity or event common to two or more events
ladder	A device for representing a set of overlapping activities

loop	An error which results in a later activity restraining an earlier activity
L.F.T.	Latest finish time of an activity
L.S.T.	Latest start time of an activity
master network	A network for the complete project
milestone	An event selected for its importance
negative total float	The time by which the duration of an activity must be reduced
network	A presentation of activities and events and their interrelationship
node	Same as event
path	An arrow or an unbroken sequence of arrows
P.N.A.	Project network analysis
P.N.T.	Project network techniques
project	An undertaking involving interrelated activities
skeleton network	A summary network with reduced activities and events but maintaining the relationships of the remaining activities and events
subnetwork	A part of a master network
summary network	A network lacking complete detail
total float	The time by which an activity may be delayed or extended without affecting the duration of the project

There is another method using diagrams known as precendence diagramming. A simple precedence diagram is shown in Fig. 6-36.

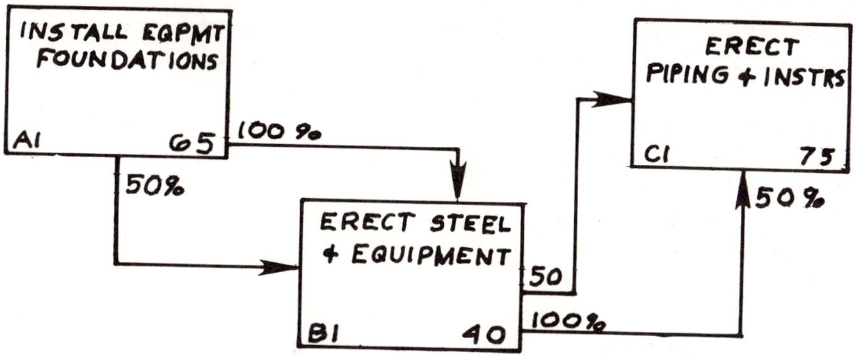

Fig. 6-36
Simple precedence diagram

Precedence diagramming uses boxes for the activities. In the Figure A1 is install equipment foundations with a duration of 65 days. When this activity is 50% completed activity B1 can be started, but B1 cannot be finished until 100% of A1 is completed. Activity B1 requires 40 days to complete. Activity B1 can be started at 50% of the time required for A1, thus at 32.5 days. When activity B1 is 50% completed activity C1 can be started. 100% of B1 must be completed before 50% of C1 can be completed. C1 requires of itself 75 days for completion.

A number of commercial computer and schedule control systems are available. They are proprietary systems for which users pay a fee. Some well known systems are

- MSCS (Management Scheduling and Control System) by McDonald Automation Co. (MCAUTO)
- Project/2 - PSDI offered by Boeing Computer Services
- PCS, PMSIV, and PROJACS by IBM
- K and H PREMIS from London, England
- TRACE by Project Management Association

- PROSYS/80 by PMI in Montreal, Quebec and marketed by Elite Data Systems, Inc.
- MISTER (Management Information System for Time Expenses and Resources) available from CSC and INFONET
- PROPLAH marketed by K and H through CDC Cybernet Services
- PACII by Information Systems, Inc., King of Prussia, Pa.
- TIMETABLE and LESSTIME marketed by UCS (United Computing Services)
- MPM (Multi-Project Management) by G. E. Information Systems
- MCS MARK II (Management Control System) by Program Control Corp., Kensington, Md.
- FASTNET by University Computing Co. (UCC)
- OPTIMA 1100 by Sperry Rand
- PERT/TIME by CDC
- PNA (Project Network Analysis) by National Cash Register Co. (NCR)
- PPS IV (Project Planning System IV) by Computing and Information Systems, Tulsa, Oklahoma
- MCS (Management Control Systems) by Construction Management Systems, Haddonfield, N.J.
- STAR (Scheduling Technique for the Allocation of Resources), by Westinghouse
- PACE and CRAM by Environmental Services, Inc., Minneapolis, Minnesota
- SPRED by CSC, El Sequndo, California
- ANGUS (A Network Guidance and Updating System) and MUSCLES, a shutdown planning system, by Union Carbide Corp.

Some Canadian-English systems are

- ASTRA II, Resource Allocation Program PERT/TIME and PERT/cost by Honeywell, Montreal
- CPM system, by G. E. Information Systems, Montreal
- DATALINE form Toronto
- PROMIS-TIME MODULE by Burroughs, Montreal
- PROSE by Concom, Montreal
- SMART I by Central Electricity Generating Board, London
- K and H from London

Some mini-computer systems available are

- Metier
- K and H PROMINI
- VISION by Systonetics
- MINIPERT by IBM

Some plotter programs available are

- EZPERT by Systonetics
- MAPS and TMAPS jointly by MCAUTO and Turner Construction Co., New York City
- K and H Project/2

7. COST INDICES AND ESCALATION

J. J. (John) O'Driscoll

Shell Oil Company, Houston, Texas

Part 1. Cost Indices

DEFINITION. A typical problem faced by an engineer involves estimating the current or future cost of some equipment, plant, or building. One way to do this is to obtain costs based on an earlier date and update these costs to the present time.

We know that prices of equipment, plants or buildings vary from time to time in accordance with competitive market conditions and the general state of inflation or deflation of a country's currency. They also vary at any given time from one area of the country to another area. These differences in price from one time to another time period and from one location to another can be measured by cost indices (also written as indexes).

A cost index is the price level at a given time and place, for a composite mix of prices, compared to the price level that existed at a base or standard time and place. Or put another way, a cost index relates the cost of an item, building or plant in terms of the cost at some particular base period and location.

There are many cost indices published by many organizations or magazines and they cover practically every area of interest to an engineer. There are indices for building construction, wage rates for various industries, types of plants or industries, and various types of equipment, material or commodities.

Remember, cost indices are based on present day costs as compared with past cost history. As published, they do not forecast future escalation, they leave that up to the individual or somebody else's discretion.

Examples of some published cost indices are:

Year	Engineering News-Record Construction	Manufacturing Wage Index	Iron and Steel	California Hwy. Dept. Highways	Chemical Engineering Plant Index
1967	100	100	100	100	100
1968	108	107	102	105	104
1969	119	113	108	111	109
1970	129	120	116	123	115
1971	148	127	123	139	121
1972	162	135	128	150	125

The above indices are based on the year 1967 as the base year. When printed by the various organizations they usually have different base years. For example, the Engineering News-Record shows two indices; one with year 1913 as base 100 and another with year 1967 as base 100. For comparison purposes, the five indices were converted to the year 1967 as base 100.

When cost indices are converted to the same base years and plotted on the same graph for comparison, the curves vary irregularly and do not correlate. This is shown in the following curve.

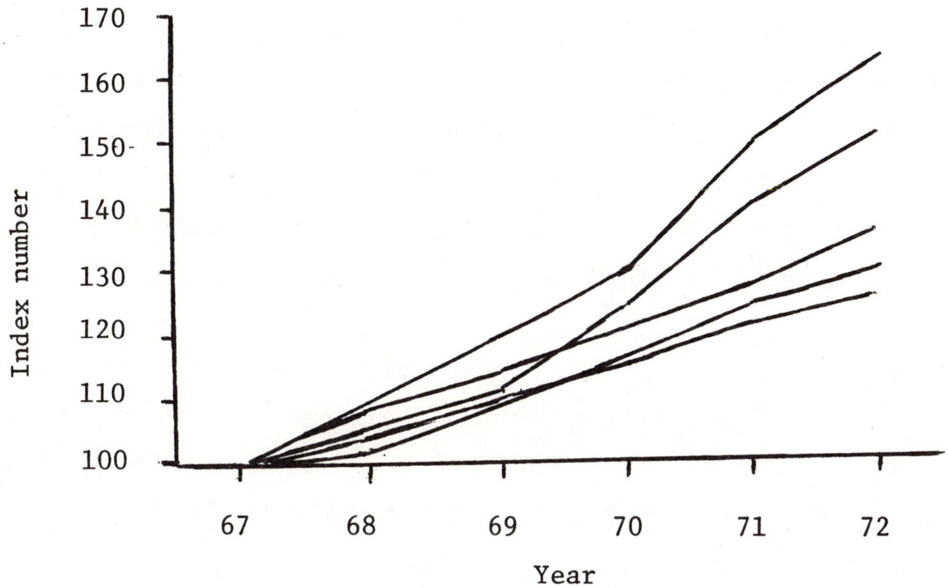

HOW TO USE AN INDEX. Cost indices are used as ratios and, therefore, should not be subtracted. To update the cost from year A to year B, multiply year's A cost by the ratio of year's B index over year's A index. Example. Using the California Highway Department's index for highways, say a highway constructed in 1969 cost \$500,000. What would the same highway cost in 1972?

$$\$500{,}000 \times \frac{150}{111} = \$676{,}000 \text{ (1972's cost)}$$

Remember that an index is an average value, and average has little meaning when applied to a specific case. Judgement has to be used as to whether the index applies to the specific cost being updated or whether the specification or scope for the new cost has changed.

HOW AN INDEX IS COMPILED. The published indices are compiled in various ways. Although they differ in nature, they have one thing in common. They are compiled for a specific piece of equipment, building, plant, or industry. They can be simple or complex. The Engineering News-Record Construction Index is simply based on the prices of certain fixed quantities of cement, lumber, structural steel, and common labor. The Marshall & Swift Index is a 47 industry average plus 12 specialized industry indices such as cement, chemical, glass, paper, petroleum, clay products, rubber, paint, electrical power equipment, mining, refrigeration, and steam power.

The Chemical Engineering chemical plant cost index is a good example of a complex index. It is made up of 67 Bureau of Labor Statistics (BLS) price indices covering 155 individual commodities, three BLS indices covering 450 additional items and BLS hourly earnings data for five groups of employees. The index also includes salary trends for engineers, draftsmen, and administrative workers. It consists of four major components:

Equipment, machinery & supports	61%
Erection & installation labor	22%
Buildings, materials and labor	7%
Engineering and supervision manpower	10%
Total	100%

The equipment, machinery and supports is made up of seven subcomponents:

Fabricated equipment	37%
Process machinery	14%
Pipe, valves and fittings	20%
Process Instruments and controls	7%
Pumps and compressors	7%
Electrical equipment and materials	5%
Structural supports, insulation & paint	10%
Total	100%

The other components are broken down to subcomponents similar to the equipment section.

To show how an index is compiled and updated, let us evaluate an index constructed for a chemical plant. For calculation of the total plant cost index, assume the components are divided into the following weighted categories; fabricated equipment, process machinery, bulk materials, construction labor and engineering. Fabricated equipment includes columns, vessels, tanks, heat exchangers and fired heaters. Process machinery includes pumps, compressors, blowers and mixers as well as instruments. Bulk materials include pipe, valves, fittings, structural steel, concrete, insulation and electrical. Construction labor includes the craftsmen working in the field. Engineering includes the design, procurement, expediting, estimating, and cost control people predominately working at the contractor's home office plus the field supervision staff, and the contractor's indirect field costs.

The percentages listed below are based on the dollars mix for say mid-1973. The index is to be calculated for July 1977 based on the relative growth of the individual mixes for each category. The July 1977 plant index calculates as follows:

	1973 Mix		July 1977 Index		
Fabricated Equipment	22%	x	293	=	64.6
Process Machinery	10%	x	213	=	21.3
Bulk Materials	24%	x	339	=	81.5
Construction Labor	28%	x	228	=	63.9
Engineering	16%	x	191	=	30.7
TOTAL	100%				262.0 (July 1977 Plant Index)

If the July 1973 was 144, the plant cost index for July 1977 indicates an increase of 81.9 percent (262 divided by 144) or a compounded average annual inflation rate of 16.1 percent for the period from mid-1973 to mid-1977.

The individual indices for each category are derived by applying to the mid-1973 values escalation factors obtained through either analysis of prices recently purchased through the company's purchasing department or through published component indices, such as published by the Bureau of Labor Statistics.

ANALYSIS OF SOME BUILDING COST INDICES. A paper by D. J. Shellenberger of Jones and Laughlin Steel Corp. printed in the Cost Engineers' Notebook published by the American Association of Cost Engineers (AACE) is used in its entirety with permission to cover this topic

"A Comparison of Building Cost Indexes
by D. J. Shellenberger
Index No. E-2.12 October, 1967

One of the important functions of cost indexes is to afford the estimator a means of upgrading the cost of a similar facility from the past to the present without going through detailed costing of individual job components. Provided the estimator uses descretion in choosing the proper index, a reasonable approximation of the present cost should result.

Only a limited amount of information has been published about many of the cost indexes which are available. A very brief description and data for most of them are published in the Quarterly "Cost Roundup" issues of the Engineering News-Record. More detailed information is available in the Cost Index

Bibliography published by the American Association of Cost Engineers. It is the purpose of this paper to present additional information about the building cost indexes and briefly to compare them and mention points to be considered when using them.

A large number of cost indexes are available to the estimator and in building construction alone, at least 15 major indexes are compiled in the United States covering from one to seventeen types of buildings in from one to over two hundred different locations. For comparison purposes I have grouped eleven of the more commonly used indexes into three categories as follows:

1. Industrial buildings in specific areas
   - Aberthaw (1 Type--New England)
   - Austin (1 Type--Central and Eastern)
   - Fruin-Colnon (5 Types--St. Louis, MO.)
2. General Buildings in specific areas
   - Fuller (Eastern cities)
   - Smith, Hinchman and Grylls (Detroit, Mich.)
3. Various types of buildings in many areas
   - American Appraisal (4 Types in 30 cities)
   - Boeckh (10 Types in 57 Areas)
   - Campbell (5 Types in 17 Areas)
   - Dow (17 Types in 237 Areas)
   - Marshall & Stevens (4 Types in Many Areas)

The other indexes are very special in nature covering railroad stations, airplane hangers, utility buildings, etc. and are not considered further.

A comparison of the first group over the period of 1935 to 1966 is presented in Figure 1. The well known Engineering News-Record Construction and Building Cost indexes are shown in the background because of their frequent use. Many fail to realize that the ENR Building Cost Index was not designed as a cost index for the construction of buildings. The two ENR indexes are

Figure 1

Comparison Of Cost Indexes

Base 1913=100

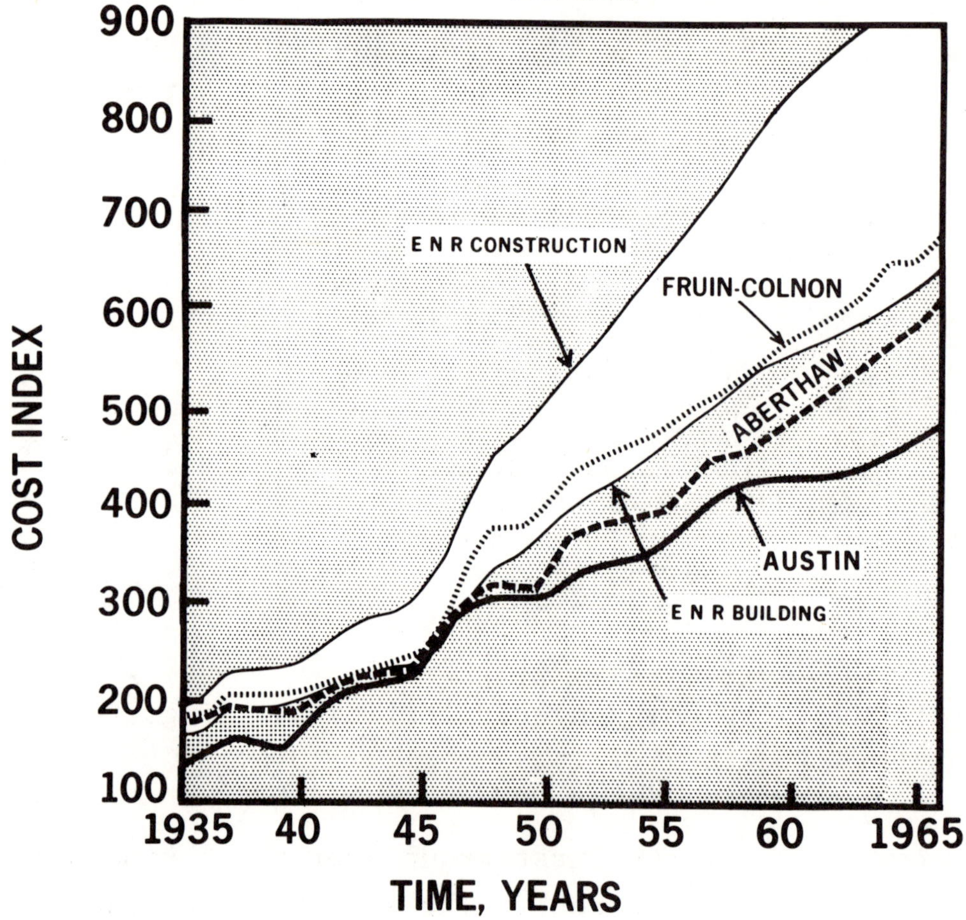

identical in that they are based on building a hypothetical block of construction. The Building Cost Index simply uses skilled labor and the Construction Cost Index uses common labor.

In all cases the figures plotted use 1913 = 100 as the base. The inflationary effect of the post World War II construction is apparent and during this period the indexes spread apart. Note that the Fruin-Colnon plot pretty much parallels the ENR Building Index on the high side whereas the Aberthaw and particularly the Austin are will below it. The differences between these indexes are the combined result of many things, but mainly they reflect differences in the rate of change of materials, labor, engineering costs between the three areas which are represented. The major components included in each of these indexes are presented in Table 1. Of the twenty separate items shown, eight of them are taken into consideration by all three companies.

Figure 2 presents a comparison of the second group. In this case, all three indexes begin and presently end at almost identical points but the path between them varies. The Fuller data follow the ENR Building Index quite closely since 1950. The Turner, and the Smith, Hinchman and Grylls indexes are close together but higher. Components included in each of these building indexes are shown in Table 2. Seven of the twenty items are taken into account by all three companies, and fourteen are considered in developing the Turner, and the Smith, Hinchman and Grylls indexes.

Figure 3 is a partial comparison of the third group using the national average figures for the American Appraisal, the Boeckh, and the Marshall & Swift indexes. These are average figures and in some cases the index for a specific area differs

COST ENGINEERS' NOTEBOOK

© 1968 AMERICAN ASSOCIATION OF COST ENGINEERS

A COMPARISON OF BUILDING COST INDEXES,
by D. J. Shellenberger,
Jones & Laughlin Steel Corporation

Index No. E-2.12
Page 4 of 9
October, 1967

Table 1

ITEMS INCLUDED IN BUILDING COST INDEXES

GROUP 1 - INDUSTRIAL BUILDINGS IN SPECIFIC AREAS

ITEM	MATERIALS				LABOR						LAND			ARCH. & ENG.						
INDEX	Cost	Sales Tax	Freight	Expediting	Cost	Efficiency	Overtime	Premiums	Procurement	FICA & Comp.	Cost	Excav. & Back.	Yard Improv.	Cost	Fee	Constr. Exp.(1)	Contractors Fee	Bidding Competition	Future Price Trends	New Concepts(2)
Aberthaw (1Type - New England)	X	—	X	X	X	X	—	—	—	X	—	X	X	—	—	X	—	X	—	X
Austin (1Type - Central & Eastern)	X	X	X	X	X	X	—	—	X	X	—	X	—	X	X	X	X	—	—	X
Fruin - Colnon (5 Types - St. Louis)	X	X	X	X	X	X	—	—	X	X	—	X	X	—	—	X	X	X	X	—

(1) Includes field supervision, field office, temporary construction · equipment rental, tools, insurance, etc.

(2) Modifications to account for new design concepts, new materials, or new construction methods.

Figure 2

Comparison Of Cost Indexes

Base 1913=100

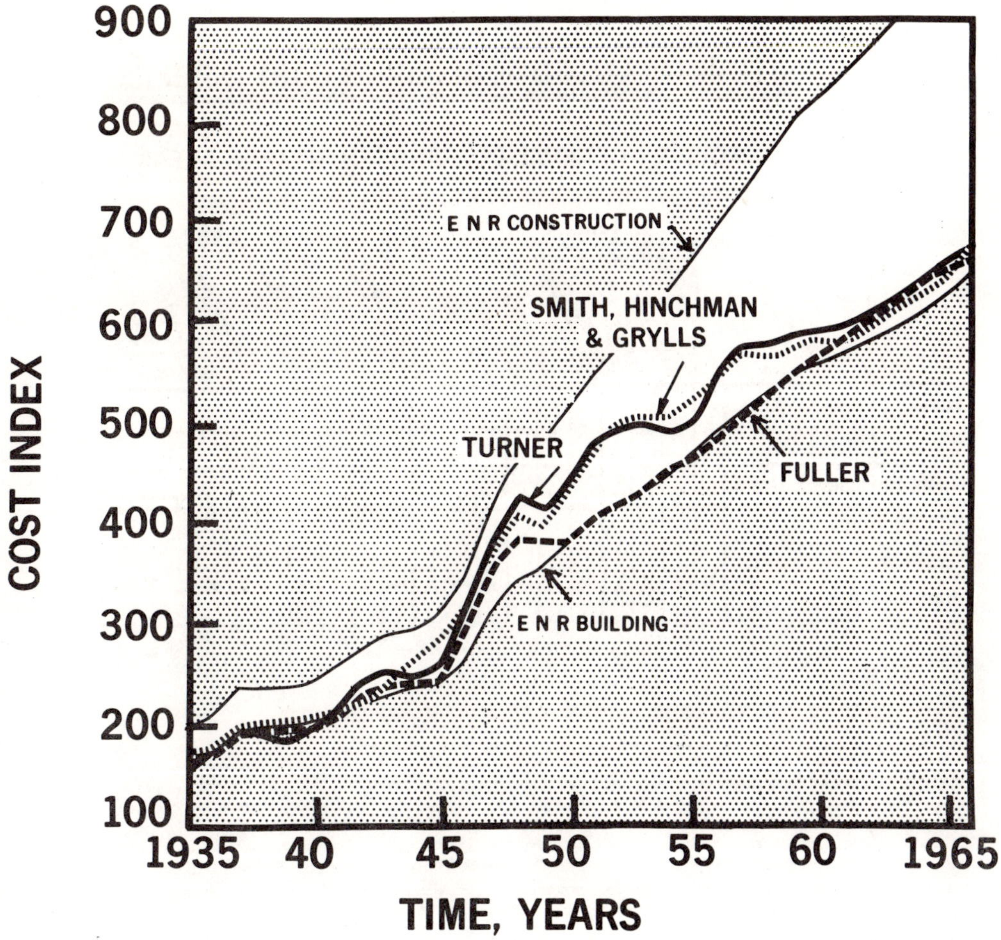

COST ENGINEERS' NOTEBOOK

© 1968 AMERICAN ASSOCIATION OF COST ENGINEERS

A COMPARISON OF BUILDING COST INDEXES,
by D. J. Shellenberger,
Jones & Laughlin Steel Corporation

Index No. E-2.12
Page 6 of 9
October, 1967

Table 2

ITEMS INCLUDED IN BUILDING COST INDEXES

GROUP 2 - GENERAL BUILDINGS IN SPECIFIC AREAS

ITEM	MATERIALS				LABOR						LAND			ARCH. & ENG.						
INDEX	Cost	Sales Tax	Freight	Expediting	Cost	Efficiency	Overtime	Premiums	Procurement	FICA & Comp.	Cost	Excav. & Back.	Yard Improv.	Cost	Fee	Constr. Exp.(1)	Contractors Fee	Bidding Competition	Future Price Trends	New Concepts(2)
Fuller (Eastern Cities)	X	X	X	—	X	X	—	—	—	X	—	X	—	—	—	X	—	—	—	X
Smith, Hinchman & Grylls (Detroit, Mich.)	X	X	X	X	X	X	X	X	X	X	—	X	X	—	—	X	X	X	X	X
Turner[3] (Eastern Cities)	X	X	X	X	X	X	—	X	—	X	—	X	X	—	—	X	X	X	X	—

(1) Includes field supervision, field office, temporary construction, equipment rental, tools, insurance, etc.

(2) Modifications to account for new design concepts, new materials, or new construction methods.

(3) The efficiency of plant and management are also considered in this index.

Figure 3

Comparison Of Cost Indexes

Base 1913=100

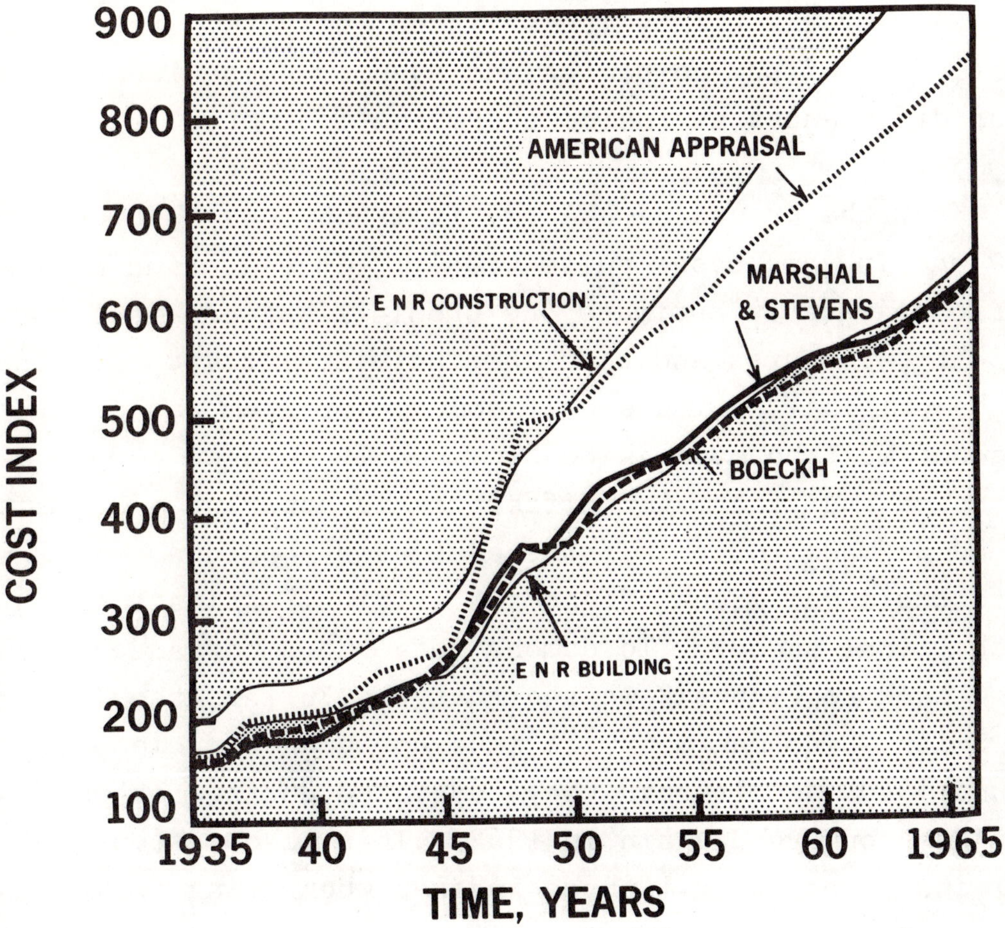

significantly from that shown. The Dow and the Campbell indexes are not included because only limited data are available and information is lacking to correct them to a 1913 = 100 base. In this case, the Boeckh, and the Marshall & Stevens indexes follow the ENR Building Index rather closely. The American Appraisal index increased much faster from 1945 to 1948 and has maintained a level substantially above the other two since that time. Components included in each of these indexes are presented in Table 3. Of the twenty items listed, only two are considered by all five of the companies and even in these items there are major differences as indicated by the footnotes.

From the preceding charts, it is obvious that the indexes vary and in some cases quite widely. In selecting a given index to upgrade an estimate one should first consider the type of building, the area and then the individual components included in the original estimate. If some major item was not included, judgment must be made as to how much it changed compared to the index and if a major difference is apparent a correction should be made for this. The estimator should not fail to appreciate the significant effect the original estimate date may have on the final number. To demonstrate the importance of this, if we go back to Figure 2 we see that the Fuller, the Turner, and the Smith, Hinchman & Grylls indexes were very close together around 1940 and although they separated during the post war period they converged again in 1962. This means essentially identical increases for all three indexes when going from 1941 (25 years ago) to 1966. However, if we were upgrading costs from 10 or 15 years back, we would find the indexes changing much differently. As shown in Table 4, if the Fuller index were used, the amount of the increase since 1951 (15 years) or 1956 (10 years) would be approximately 60% more than the increase indicated by the Turner, or the Smith, Hinchman & Grylls indexes.

(End of quotation from Shellenberger)

Table 3

ITEMS INCLUDED IN BUILDING COST INDEXES

GROUP 3 - VARIOUS TYPES OF BUILDINGS IN MANY AREAS

ITEM	MATERIALS				LABOR						LAND			ARCH. & ENG.						
INDEX	Cost	Sales Tax	Freight	Expediting	Cost	Efficiency	Overtime	Premiums	Procurement	FICA & Comp.	Cost	Excav. & Back.	Yard Improv.	Cost	Fee	Constr. Exp.(1)	Contractors Fee	Bidding Competition	Future Price Trends	New Concepts(2)
American Appraisal (4 Types - 30 Cities)	(3) X	—	X	—	(3) X	X	—	—	—	—	—	—	—	—	—	X	X	—	—	—
Boeckh (10 Types - 57 Areas)	X	X	—	—	X	X	X	X	X	X	—	—	—	X	X	X	X	—	—	—
Campbell (5Types - 17 Areas)	(3) X	—	—	—	(3) X	—	—	X	—	—	—	—	—	—	—	—	—	—	—	X
Dow (F. W. Dodge) (17 Types - 237 Areas)	(4) X	—	X	—	(4) X	—	—	—	—	X	—	—	—	—	—	—	—	X	—	—
Marshall & Stevens (4 Types - Many Areas)	X	—	—	—	X	—	—	—	—	—	—	—	—	—	—	—	—	—	—	X

(1) Includes field supervision, field office, temporary construction, equipment rental, tools, insurance, etc.

(2) Modifications to account for new design concepts, new materials, or new construction methods.

(3) Building services (lighting, heating, ventilation, plumbing, etc.) are not included.

(4) A standard group of materials and labor rates are used.

Table 4

Effect of Original Estimate Date on Upgraded Cost for Different Indexes

Date Of Original Estimate	% Increase in Index Thru 1966			
	1941 (25 Yrs.)	1951 (15 Yrs.)	1956 (10 Yrs.)	1961 (5 Yrs.)
Fuller	225	65	39	16
Smith, Hinchman & Grylls	213	43	25	18
Turner	212	41	24	14

LIMITATIONS IN USING PUBLISHED INDICES. There is no perfect index. While a large amount of time may go into publishing an Index, it has limitations.

1. A cost index is a statistical average and suffers from all the disadvantages of averages. While the average cost for installing concrete may be $150 per cubic yard, this does not mean that any one foundation will cost exactly $150 per cubic yard.
2. Most indexes represent reproduction costs. They do not account for any radical technological changes in design or construction. For example, take what happened to the design of warehouses over the past ten or twenty years. Today's metal frame construction with aluminum and transite skins and metal pan roof decks have replaced heavier masonry, concrete, wood and steel construction. A weighting based on the mix of materials twenty years ago will hardly be representative of today's construction, not just in terms of material and labor ratios, but in terms of the materials themselves. Therefore, any fixed base weight index is doomed to obsolescence in today's rapidly changing technology. Periodic re-study of the index components and mix is required.
3. There is a reporting time lag. The time required to collect, analyze, and publish the information takes several months. This is especially true when an index is based in part on indices published by other organizations. For example, the BLS is undoubtly the primary national source of information and is used in many indices to a greater of lesser degree. Yet the BLS index or data runs two months late. In summary, most indices are about three to six months behind the

time. One exception to this is the Engineering News-Record's Construction Index. Because of its simplicity, pricing of the three basic commodities plus labor is on a current basis.

·4 They lack sensitivity to short term economic cycle swings. Most indices are based on 1) published list prices rather than actual market prices; and 2) average labor conditions rather than actual conditions. The realistic index will appear somewhat like a roller coaster superimposed on a long term creeping inflation trend shown here as:

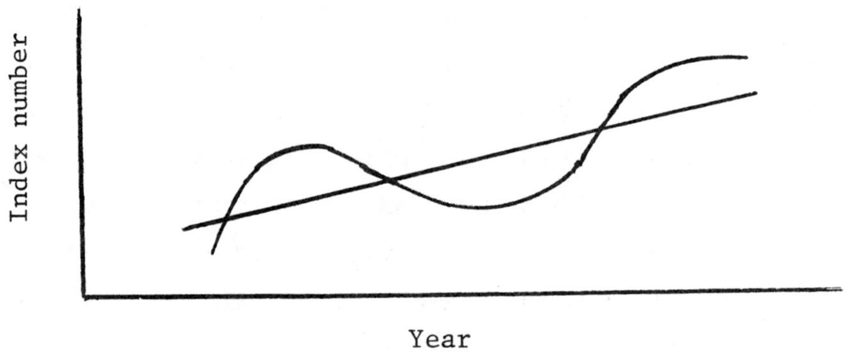

There are several significant variables contributing to abnormal increases. In a period of rising construction employment, labor productivity drops as marginal workers enter. When the bottom of the barrel is reached, overtime is used with its double time premiums. Then with labor in tight supply, wage rate contract negotiations become more difficult, and abnormally high wage concessions may result. The same is true for materials. As the economy is booming, material costs escalate more rapidly as the economy moves into a seller's market. List prices may not move up but, more importantly, discounts to aggressive purchasing departments disappear. Plus premium costs are used to insure deliveries in reasonable time schedules. As the boom peaks and the trend reverses, costs of construction actually drop - even though it may not show up in indices. The base labor wage rates do not reduce, but as overtime disappears and labor productivity returns to average, or better than average, absolute labor costs decrease. Equally, the return to tight competitive bidding conditions by material suppliers causes material cost reductions. If the market gets tight enough, vendors bid either work or equipment at out of pocket costs with very little or no profit margin. Then the cycle starts all over again.

Part 2. Escalation

INTRODUCTION. As stated earlier, most published indices do not forecast future escalation. The Engineering News-Record's indices do forecast the escalation for the following year. They do not forecast beyond one year.

A simple way to guess escalation is to plot cost index trends on past history and then extrapolate into the future. This requires a considerable degree of judgment on forecasting escalation for construction projects. Some companies try to gain insight into the movements of a construction cost index by comparing its history to broad price indices for the entire economy. Some broad price indices are: a) Gross National Product Deflator; b) Plant & Equipment Deflator; c) Wholesale Price Index; and d) Comsumer Price Index. These indices are maintained by the Bureau of Labor Statistics.

Periodically, government economists predict escalation

rates for these broad price indices. These predictions can be applicable to construction cost indices.

PROCEDURE FOR ESTIMATING ESCALATION. It is not advisable to apply escalation to the overall Index. A good approach is to analyze the movements of components of the index to be forecasted. Then, based on applying escalation rates to the various components, an overall composite escalation can be calculated.

A suggested procedure is

1. Determine the percentage cost breakdown for the components of the index. A breakdown for the guardhouse project is: a) construction labor; b) materials; and c) subcontractors.
2. Determine the rate of escalation for each of the components. This can be based either on judgment or published forecasts.
3. Determine the schedule for the construction projects. This perhaps is the key to this approach. The calculation is meaningless unless a reasonable judgment is made of the time interval applicable to each component. Each component does not start and finish at the same time. After the design is well underway, the material components are purchased. Construction labor and subcontractors start after the material is delivered to the jobsite. A similar sequence follows for the finish of the components.
4. Computation.
    - a) The total estimated cost is $7,166 - based on today's costs.
    - b) Breakdown of the components - material 56%; construction labor 35%; and subcontractors 9%.
    - c) Project is to start three (3) years from now.
    - d) Time schedule to mid-point of schedule - materials $\frac{1}{4}$ year, labor $\frac{1}{2}$ year, and subtractors $\frac{3}{4}$ year.
    - e) Rate of escalation - materials 6%/year, labor 11%/year, and subcontractors 10%/year.

The total project cost as calculated is

·Material:	$\$7,166 \times 56\% \times (1.06)^{3.25}$	= \$ 4,850
·Labor:	$\$7,166 \times 35\% \times (1.11)^{3.5}$	= \$ 3,614
·Subcontractors:	$\$7,166 \times 9\% \times (1.10)^{3.75}$	= \$ 922
	Total Estimated Project Cost	\$ 9,386

The escalation added to the basic estimate is 9,386 - 7,166 = \$2,220.

Note that the rates have been compounded in calculating the escalation. The effect of compounding is relatively insignificant for short time periods and errors in the assumed escalation rates and breakdown could introduce greater differences. For delayed projects compounding can be significant, especially if several years are involved. For this example, if the escalation rates are not compounded, the escalation is \$1,990 or 10% less than the compounded escalation.

Part 3. Types of Contracts

FOUR MAJOR TYPES can be recognized. An enumeration with variations follows

1. Cost-plus contracts
   - Percentage fee, augmented perhaps with a maximum fee
   - Fixed fee
   - Time and material
2. Incentive contracts
   - Sliding scale fee
   - Guarantee maximum with sharing of cost savings
   - Time of completion--bonus or penalty
3. Lump sum contracts
   - Based on drawings and specifications
   - Convertible from cost plus to lump sum
4. Unit price contracts
   - Flat rate
   - Sliding Scale

REFERENCES

1. Jelen, F. C., Cost and Optimization Engineering , McGraw-Hill Book Company, 1970.

2. Hackney, J. W. , Control and Management of Capital Projects , John Wiley & Sons, Inc., 1965.

3. Norden, R. B., "Development of a Cost Index for Plants", AACE Transactions, Houston, 1968.

4. Patterson, W. H., "Preparing and Maintaining a Construction Cost Index", AACE Transactions, Pittsburgh, 1969.

5. Shellenberger, D. J., "A Comparison of Building Cost Indexes", AACE Cost Engineers' Notebook. Index No. E-2.12. October 1967.

6. Caldwell, D. W., "Forecasting Capital Project Escalation in the Chemical Industry", AACE Transactions, Orlando, 1975.

7. Jones, L. R., "Estimating Cost Escalation for Refinery Capital Projects", AACE Transactions, San Francisco, 1970.

8. MANAGEMENT SCIENCE TECHNIQUES

Jerry L. Hamlin

Cities Service Company, Tulsa, Oklahoma

OUTLINE. An outline for the chapter follows

I. Operations research
II. Linear Programming (LP)
    A. Introduction
        Manufacturing problem
        Product-mix problem
        Investment problem
    B. Two dimension graphical solution
    C. The Simplex method for solving linear programs
    D. Conversion of LP problems to the Simplex form
    E. Interpretation of Simplex solutions
III. Dynamic programming (DP)
    A. Introduction
        Stagecoach problem
    B. Characteristics of dynamic programming problems
    C. Other examples of DP problems
        Product allocation problem
        Probability of success problem
        Integer LP problem
    D. Discussion

IV. Simulation
   A. Introduction
      Coin problem
   B. Random variables
   C. Additional simulation examples
      Queueing problem
   D. Discussion

OPERATIONS RESEARCH may be generally described as a scientific approach to decision making that involves the operations of organizational systems. It is widely used in industry. The most common and useful techniques are linear programming, dynamic programming, game theory, network analysis (including PERT and CPM), queueing theory, and simulation techniques.

Operations research involves many problems in optimization and thus has wide application in cost engineering where minimum cost or maximum profit are prime considerations. The present chapter will aid the reader in understanding the techniques used in linear programming, dynamic programming, and simulation. All are very useful in the management science aspects of cost engineering.

Part 1. Linear Programming (LP)

INTRODUCTION. Linear programming deals with the problems of allocating limited resources among competing activities in some optimal manner. All the mathematical functions in the LP model must be linear.

Linear programming is concerned with a solution (where it exists) to problems of the form: Optimize (i.e, maximize or minimize)

$$z = k_1x_2 + k_2x_x + \dots + k_Jx_J \qquad (1)$$

subject to the constraints

$$\begin{array}{lccccl} \text{I.} & a_{11}x_1 & + a_{12}x_2 & + \dots & + a_{1J}x_J & \geq a_{10} \\ & \vdots & \vdots & & \vdots & \vdots \\ & a_{A1}x_1 & + a_{A2}x_2 & + \dots & + a_{AJ}x_J & \geq a_{A0} \end{array}$$

$$\text{II.}\quad b_{11}x_1 + b_{12}x_2 + \cdots + b_{1J}x_J \leq b_{10}$$
$$\vdots \qquad \vdots \qquad\qquad \vdots \qquad \vdots$$
$$b_{B1}x_1 + b_{B2}x_2 + \cdots + b_{BJ}x_J \leq b_{B0}$$

$$\text{III.}\quad c_{11}x_1 + c_{12}x_2 + \cdots + c_{1J}x_J = c_{10}$$
$$\vdots \qquad \vdots \qquad\qquad \vdots \qquad \vdots$$
$$c_{C1}x_1 + c_{C2}x_2 + \cdots + c_{CJ}x_J = c_{C0}$$

Thus, we have a *linear* objective function which is to be optimized subject to a total of (A + B + C) *linear* constraints involving inequalities ($\leq$, $\geq$) and equalities.

One key to the successful utilization of linear-programming technology is the ability to recognize when a problem can be solved by linear programming and to subsequently formulate the mathematical model. Let us examine some illustrative models.

MANUFACTURING PROBLEM. A manufacturing firm has excess capacity to devote to one or more of three products called 1, 2, and 3. The excess capacity is

(1)	Milling machine	200 hrs/wk	A
(2)	Lathe	100 hrs/wk	B
(3)	Grinder	50 hrs/wk	C

The manufacture of the three products requires the following machine time (in hours/unit product):

(1)	Product 1;	8 milling mac.,	4 lathe,	2 grinder	20 40 50 } 10
(2)	Product 2;	2 " " ,	3 " ,	0 "	40 10 5 } 15
(3)	Product 3;	3 " " ,	0 " ,	1 "	

The sales potential for products 1 and 2 is estimated to exceed the maximum production rate and for product 3 it is estimated to be 20 units per week.

The unit profit would be \$20, \$6, and \$8, respectively, on products 1, 2, and 3. What production schedule will maximize the profit?

Let x_i $(i = 1,2,3)$ be the number of units of product i produced per week. The measure of effectiveness is the profit, denoted by z; we wish to maximize it.

The mathematical formulation is maximize

$$z = 20x_1 + 6x_2 + 8x_3$$

subject to the restrictions

$$\text{Machine capacity} \begin{cases} 8x_1 + 2x_2 + 3x_3 \leq 200 \\ 4x_1 + 3x_2 \leq 100 \\ 2x_1 + x_3 \leq 50 \end{cases}$$

$$\text{Sales potential} \quad x_3 \leq 20$$

$$\text{Non-negativity} \quad x_1, x_2, x_3 \geq 0$$

PRODUCT MIX PROBLEM. A refinery wishes to blend four petroleum constituents, called constituents 1, 2, 3, and 4, into three grades of gasoline, called products A, B, and C. The constituent availability (bbls/day) and cost per barrel are

(1) Constituent 1; 3,000 @ \$3
(2) Constituent 2; 2,000 @ \$6
(3) Constituent 3; 4,000 @ \$4
(4) Constituent 4; 1,000 @ \$5

The mixtures of the gasoline products are:

(1) Product A; $\leq$ 30% of 1, $\geq$ 40% of 2 and $\leq$ 50% of 3
(2) Product B; $\leq$ 50% of 1 and $\geq$ 10% of 2
(3) Product C; $\leq$ 70% of 3

The selling price per barrel is \$5.50, \$4.50, and \$3.50, respectively, for products A, B, and C. If there are no sales potential or equipment availability considerations, what product mix will maximize the profit?

Let x_{A1} be the amount of constituent 1 allocated to product A, x_{B2} be the amount of constituent 2 allocated to product B, etc. Denoting the profit by z, we wish to maximize

$$\begin{aligned} z &= (\text{total sales}) - (\text{total costs}) \\ &= 5.5(x_{A1} + x_{A2} + x_{A3} + x_{A4}) + 4.5(x_{B1} + x_{B2} + x_{B3} + x_{B4}) \\ &\quad + 3.5(x_{C1} + x_{C2} + x_{C3} + x_{C4}) - 3(x_{A1} + x_{B1} + x_{C1}) \\ &\quad - 6(x_{A2} + x_{B2} + x_{C2}) - 4(x_{A3} + x_{B3} + x_{C3}) - 5(x_{A4} + x_{B4} + x_{C4}) \end{aligned}$$

subject to the restrictions

Availability
$$\begin{cases} x_{A1} + x_{B1} + x_{C1} \le 3000 \\ x_{A2} + x_{B2} + x_{C2} \le 2000 \\ x_{A3} + x_{B3} + x_{C3} \le 4000 \\ x_{A4} + x_{B4} + x_{C4} \le 1000 \end{cases}$$

(Blend restrictions): for Product A
$$\begin{cases} x_{A1} \le 0.3(x_{A1} + x_{A2} + x_{A3} + x_{A4}) \\ x_{A2} \ge 0.4(x_{A1} + x_{A2} + x_{A3} + x_{A4}) \\ x_{A3} \le 0.5(x_{A1} + x_{A2} + x_{A3} + x_{A4}) \end{cases}$$

for Product B
$$\begin{cases} x_{B1} \le 0.5(x_{B1} + x_{B2} + x_{B3} + x_{B4}) \\ x_{B2} \ge 0.1(x_{B1} + x_{B2} + x_{B3} + x_{B4}) \end{cases}$$

for Product C
$$x_{C1} \le 0.7(x_{C1} + x_{C2} + x_{C3} + x_{C4})$$

Non-negativity
$$x_{A1}, \ldots, x_{A4}, x_{B1}, \ldots, x_{B4}, x_{C1}, \ldots, x_{C4} \ge 0.$$

INVESTMENT PROBLEM. An investor has \$10,000 which can be invested in money-making activities A, B, C, and D. The money-making activities become available as follows:

(1) A and B; immediately
(2) C; beginning of second year
(3) D; beginning of fifth year

The profit payouts for each dollar invested in a money-making activity are the following:

(1) A; $0.40 after 2 years
(2) B; $0.70 after 3 years
(3) C; $1.00 after 4 years
(4) D; $0.30 after 1 year

Assuming that each investment can be immediately renewed after a payout period, how should the $10,000 be invested in order that the total accumulated payout be maximized at the end of 5 years?

Let z represent the total payout after 5 years, and 1st x_{A1} represent the amount in dollars invested in activity A the first year, etc. The problem is then to maximize

$$z = 0.4(x_{A1} + x_{A2} + x_{A3} + x_{A4}) + 0.7(x_{B1} + x_{B2} + x_{B3}) + 1.0x_{C2} + 0.3x_{D5}$$

subject to the restrictions

Year 1 $x_{A1} + x_{B1} \le 10{,}000$

Year 2 $x_{A1} + x_{B1} + x_{A2} + x_{B2} + x_{C2} \le 10{,}000$

Year 3 $x_{B1} + x_{A2} + x_{B2} + x_{C2} + x_{A3} + x_{B3} \le 10{,}000 + 0.4x_{A1}$

Year 4 $x_{B2} + x_{C2} + x_{A3} + x_{B3} + x_{A4} \le 10{,}000 + 0.4(x_{A1} + x_{A2}) + 0.7x_{B1}$

Year 5 $x_{C2} + x_{B3} + x_{A4} + x_{D5} \le 10{,}000 + 0.4(x_{A1} + x_{A2} + x_{A3}) + 0.7(x_{B1} + x_{B2})$

Non-negativity $x_{A1}, \ldots, x_{A4}, x_{B1}, \ldots, x_{B3}, x_{C2}, x_{D5} \ge 0$

Each of the preceding sample problems contained three or more variables. Where no more than three variables are involved, a graphical solution can be obtained. A two-variable problem involves a 2 D display and can best be utilized to demonstrate the technique.

TWO DIMENSIONAL GRAPHICAL SOLUTION. The profits from products 1 and 2 are \$1 and \$2, respectively, per unit. Assuming that x_1 and x_2 units of the products are to be produced, we wish to maximize the profit

$$z = 1(x_1) + 2(x_2)$$

subject to the restrictions -

(1) $-x_1 + 3x_2 \leq 10^4$

(2) $x_1 + x_2 \leq 6(10^3)$

(3) $x_1 - x_2 \leq 2(10^3)$

(4) $x_1 + 3x_2 \geq 6(10^3)$

$x_1, x_2 \geq 0$

The figure on the next page shows pictorally the restricted solution area and the straight-line loci for two values of the profit within the area; We note first that the non-negativity constraint eliminates other than the first quadrant from consideration. Next, we note that as the profit z increases, its straight-line representation progresses upward and towards the right at a slope of 2, and the loci of constant z values are parallel, with a slope -0.5. Clearly the maximum value of z which satisfies the restrictions occurs at the intersection of restriction boundaries (1) and (2); it is $z = 10^4$, for which $x_1 = 2(10^3)$ and $x_2 = 4(10^3)$.

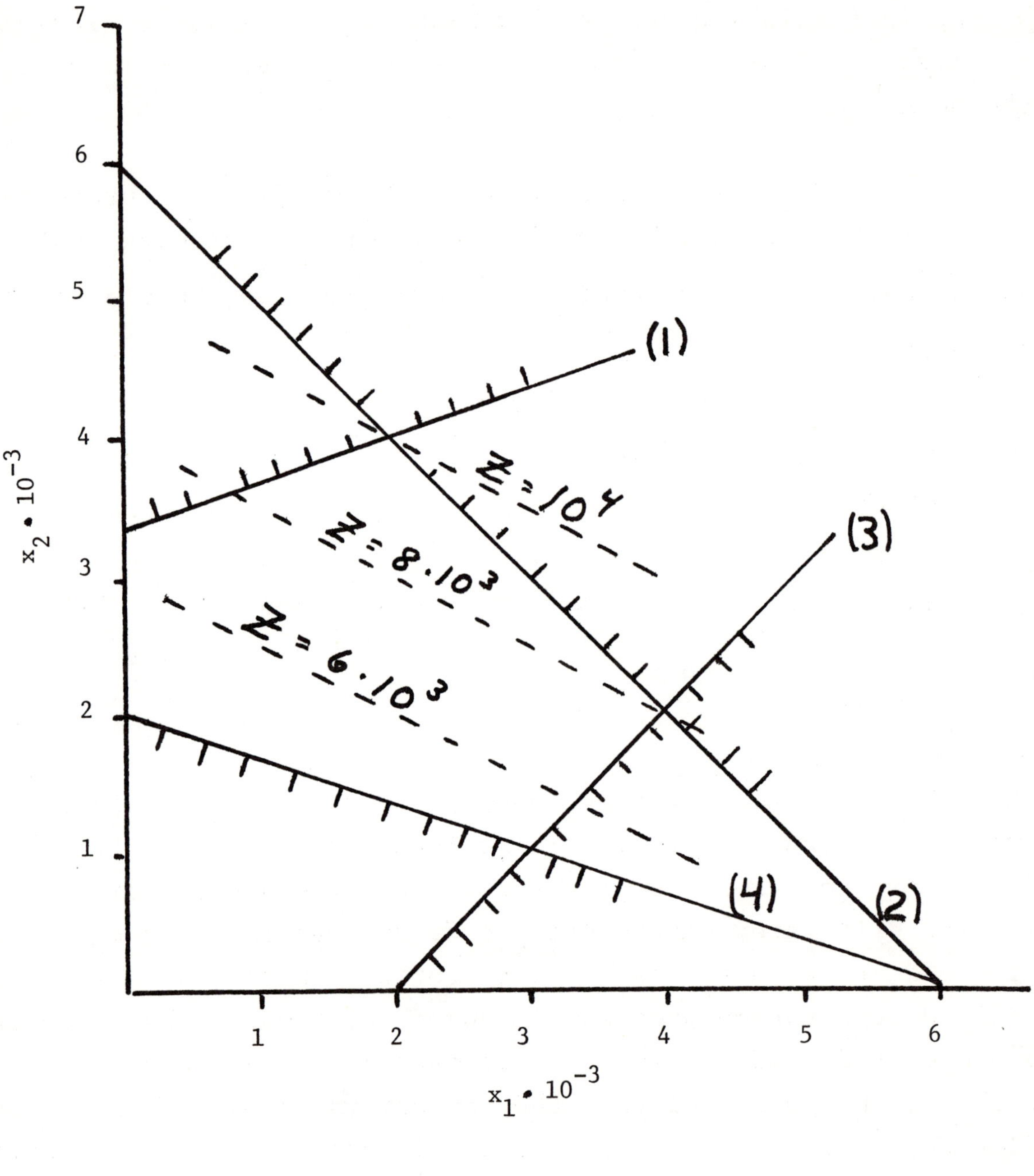
7
6
5
4
3
2
1
1
2
3
4
5
6
$x_2 \cdot 10^{-3}$
$x_1 \cdot 10^{-3}$
(1)
(2)
(3)
(4)
Z = 10^4
Z = $8 \cdot 10^3$
Z = $6 \cdot 10^3$

The solution is unique and it is finite. Had restriction (2) not been imposed, the solution would have occurred at the intersection of restriction boundaries (1) and (3). Had both restrictions (1) and (2) not been imposed, there would have been no finite solution for z. Had the profit been given by $z = x_1 + x_2$, we see that there would have been a finite line of solutions along restriction boundary (2) between its intersections with restriction boundaries (1) and (3).

SIMPLEX METHOD. Any value $(x_1, x_2, \ldots, x_N)$ of the linear-programming variables for which all of the restrictions are satisfied is called a "feasible point or solution." The function z to be optimized is called the "objective function," and a feasible point which maximizes it is called an "optimal solution." The intersections of the restriction boundaries on the perimeter of the feasible solution region are called "extreme feasible points or solutions." The optimal solution can be shown to always be an extreme feasible point, due to the linearity of the objective function and the restrictions; in case two extreme points are solutions, then the optimal solution is "nonunique" and it consists of the entire edge joining the points.

Since the solution occurs at an extreme feasible point, we may begin at some such point and compute the value of the objective function z. We then exchange this extreme point for its counterpoint at the other end of an edge in such a manner that a larger value of z is obtained; keep in mind that an increasing number of edges will intersect at any extreme feasible point as dimension of the linear-programming problem increases. The process of exchange and edge following continues until z can no longer be increased, at which point (or points) we have an optimal solution. In this manner, all of the feasible points are not examined, and their number tends to be prohibitive for large problems. Furthermore, since each successive extreme point must increase the value of z, it is impossible to return to an extreme point previously reached. A well-known algorithm for effecting this procedure is called the "simplex" method.

It is beyond the scope of this chapter to investigate the details of the simplex algorithm. Rather, we will state precisely the form of the problem which the method will solve, and concentrate on converting linear-programming problems into that form. The problem can then easily be formulated for and solved by a computer program.

The simplex algorithm solves problems of the form: Maximize

$$z = c_1x_1 + c_2x_2 + \ldots + c_Nx_N \qquad (2)$$

subject to the constraints

$$\begin{array}{lll} (1) & a_{11}x_1 + a_{11}x_2 + \ldots + x_{N-M+1} & = b_1 \\ (2) & a_{21}x_1 + a_{22}x_2 + \ldots + 0 + x_{N-M+2} & = b_2 \\ & \vdots & \\ (M) & a_{M1}x_1 + a_{M2}x_2 + \ldots + 0 + 0 + \ldots + x_N & = b_M \end{array}$$

$$x_1, x_2, \ldots, x_N, b_1, b_2, \ldots, b_M \geq 0.$$

We note that each of the variables $x_{N-M+1} \ldots x_N$ occurs uniquely in a single constraint, each having a coefficient value of 1.0. They are termed the "initial basis variables."

The above problem differs considerably from the previously stated problems, expression (1), with which linear programming is concerned. We will proceed to examine the plausibility of converting the latter problems into the simplex form.

CONVERSION TO SIMPLEX FORM. The conversion to the simplex form is accomplished by the introduction of "slack," "surplus," and "artificial" variables, in addition to some mathematical finagling. We have the following definitions:

- *slack variable* - the positive amount added to the left-hand side of a "$\leq$" constraint to give an equality "=".
- *surplus variable* - the positive amount subtracted from the left-hand side of a "$\geq$" constraint to give an equality "=".
- *artificial variable* - a variable which is introduced into the linear programming problem in order to aid in the conversion to the simplex form, but which will be forced to be zero in the optimal solution.

Let us consider the conversion of a simple linear-programming problem to the simplex form through the utilization of only slack and surplus variables, e.g.,

Original Form	Simplex Form
Maximize: $z = 2x_1 + x_2$	Maximize: $z = 2x_1 + x_2 + 0(x_3 + x_4)$
Subject to the restrictions	
(1) $x_1 + x_2 \leq 6$	(1) $x_1 + x_2 + x_3 = 6$
(2) $x_1 + 3x_2 \geq -5$	(2) $-x_1 - 3x_2 + x_4 = 5$
$x_1, x_2 \geq 0$	$x_1, \ldots x_4 \geq 0$

Here, we have introduced as as-yet-unknown slack variable x_3, which simply takes up the slack between the two sides of the "$\leq$" sign and, likewise, an as-yet-unknown surplus variable x_4 has been introduced. It is clearly seen that as long as the slack variable satisfies the single restriction, $x_3 \geq 0$, that $(x_1 + x_2) \leq 6$ in the equation $(x_1 + x_2) + x_3 = 6$; a similar argument applies to restriction (2). Upon examining the basic simplex form of the linear-programming problem, we see that x_3

and x_4 have become the initial basis variables and, importantly, the right-hand constants of the simplex form of the restrictions are positive, as they must be.

The above example was quite simple in that there were no complications in its conversion. Let us consider the various complications and their solutions (numbered 1 to 4).

1. Non-positive b_i; Had constraint (2) of the previous problem been $x_1 + 3x_2 \geq 5$, then the simple introduction of the surplus variable x_4 leads to the equivalent simplex expression $-x_1 - 3x_2 + x_4 = -5$. In order to make the right-hand constant positive, we can rewrite the expression as $x_1 + 3x_2 - x_4 = 5$, but now x_4 has a -1.0 coefficient and does not meet the criteria for a basis variable. We can introduce a fifth positive variable x_5 to be added to the left hand side of this expression which will serve as an initial basis variable for the expression, but the original inequality (2) will no longer be represented by the expression unless x_5 is forced to zero. We can drive this "artificial" variable to zero by including it in the original objective function with a large negative coefficient, say (-M). The simplex formulation of the entire sample problem then becomes maximize

$$z = 2x_1 + x_2 - Mx_5 (M>>0)$$

subject to the restrictions

$$(1)\ x_1 + x_2 + x_3 = 6$$

$$(2)\ x_1 + 3x_2 - x_4 + x_5 = 5$$

$$x_1, \ldots\ldots, x_5 \geq 0$$

We see that since $x_5 \geq 0$, the objective function can only be maximized when $x_5 = 0$. This is commonly called the "Big M" method.

·2. Equality in the restrictions. The problem here is that there is no evident means to introduce a slack or surplus variable in order that an equality contain an initial basis variable. One method for alleviating this problem is to replace an equality by a pair of inequalities, e.g.,

$$3x_1 + 2x_2 = 18$$

is equivalent to the pair of inequalities

$$3x_1 + 2x_2 \leq 18$$
$$3x_1 + 2x_2 \geq 18$$

When there is more than one equality, we can add all the expressions together and then convert the composite expression into a pair of inequalities. Perhaps the simplest and most desirable technique for alleviating the equality restriction problem is simply to introduce an artificial variable in the left-hand side of the equality and include it also with a large negative coefficient in the objective function, i.e., we apply the Big M method.

·3. Variables unconstrained in sign. We note that all variables in the simplex form of the linear-programming problem are constrained to be ≥ 0. Certainly many realistic linear-programming problems can be developed in which the variables must be negative in the optimum solution. We solve the problem in such cases by merely replacing each unconstrained variable, say x_i, by the simple difference in two positive variables.

Thus, we make the substitution

$$x_i = x_i' - x_j' \qquad (x_i, x_j \geq 0),$$

where x_j' is a new variable, i.e., the subscript j has not been previously assigned in the problem. We must then remember to convert back to the original variable x_i by subtracting x_j from it (x_i') in the optimum solution. This procedure is implemented independently for each such unconstrained variable.

4. Objective function to be minimized. Here we simple utilize the principle that minimizing any function $f(x_1, x_2, \ldots, x_n)$ subject to a set of restrictions is completely equivalent to maximizing $-f(x_1, x_2, \ldots, x_n)$ subject to the same restrictions. We then take the negative of the optimum solution for the latter objective function to obtain the minimized result.

Let us consider the conversion of a second linear-programming problem to the simplex form, where all of the preceding complications exist. The original form of the problem is minimize

$$z = 2x_1 + x_2 - 6x_3 - x_4$$

subject to the Restrictions -

(1) $3x_1 + x_4 \leq 25$

(2) $x_1 + x_2 + x_3 + x_4 = 20$

(3) $4x_1 + 6x_3 \geq 5$

(4) $2 \leq 2x_2 + 3x_3 + 2x_4 \leq 30$

(5) $2x_1 + x_2 - 6x_3 - x_4 \leq 100$

$x_1, \ldots, x_3 \geq 0$

The simplex form of the problem is maximize

$$(-z) = -\left[2x_1 + x_2 - 6x_3 - (x_4' - x_5')\right] - M(x_7 + x_9 + x_{11}$$

subject to the restrictions

$$
\begin{aligned}
&(1)\ 3x_1 + (x_4'-x_5') + x_6 = 25\\
&(2)\ x_1 + x_2 + x_3 + (x_4'-x_5') + x_7 = 20\\
&(3)\ 4x_1 + 6x_3 - x_8 + x_9 = 5\\
&(4)\ 2x_2 + 3x_3 + 2(x_4'-x_5') - x_{10} + x_{11} = 2\\
&(4')\ 2x_2 + 3x_3 + 2(x_4'-x_5') + x_{12} = 30\\
&(5)\ 2x_1 + x_2 - 6x_3 - (x_4'-x_5') + x_{13} = 100
\end{aligned}
$$

$$x_1, \ldots\ldots, x_3, x_4', x_5', x_6, \ldots\ldots, x_{13} \geq 0$$

The solution to the simplex formulation of the problem is

$$(x_1, \ldots\ldots, x_3, x_4, x_5, x_6, \ldots\ldots, x_{13}; -z) =$$

$$(20,5,30,0,35,0,0,255,0,28,0,0,0,0,;100)$$

We now must make the conversion $x_4 = x_4' - x_5'$ and take the negative of the objective function to obtain the solution to the original problem, i.e., we obtain

$$(x_1, \ldots\ldots, x_4; z) = (20,5,30,-35;-100),$$

where ($z = 2x_1 + x_2 - 6x_3 - x_4$) has been minimized, subject to the specified restrictions.

INTERPRETATION OF SIMPLEX SOLUTIONS. There are several outcomes to a simplex solution of a linear-programming problem, vis., 1) a unique solution is obtained, 2) there are multiple solutions (nonunique solution), 3) no feasible solution exists, and 4) no bounded solution exists. A discussion of the last three outcomes follows.

Multiple Solutions. There are no means for analyzing the uniqueness of a single simplex solution; and, if there are multiple solutions, the determination of all solutions requires a special algorithm which is extremely sensitive to computer round-off error. However, the simplex problem can be reformulated with the subscripts of the variables in the original objective function reversed in order; a different solution should be obtained if the original solution was not unique. Where there are multiple solutions, an infinite number of optimal solutions can be constructed by taking weighted averages of the simplex solutions; thus, if the two solutions $(x_1,x_2) = (2,3)$ and $(3,4)$ are obtained, then other optimal solutions are, e.g., $(x_1,x_2) = (2/2 + 3/2,\ 3/2 + 4/2)$, $(2/3 + 6/3,\ 3/3 + 8/3)$, etc.

·No Feasible Solution. If one of the artificial variables has a value other than zero in the final solution, then no feasible solution exists for that particular simplex formulation. The coefficient M of the artificial variables in the objective function can be increased (providing that rould-off error is not prohibitive) and another solution attempted, as a further check on the existence of a feasible solution.

·Unbounded Solution. Clearly, not all linear-programming problems will have finite or bounded solutions. There are definite checks for the existence of this situation in the simplex algorithm, and most simplex computer programs will indicate this as a result when it occurs.

An additional feature of the simplex solution is that redundancies in the constraints will result in fewer non-zero variables in the final simplex solution than there are constraints, although this occurrence does not insure that there are reduncancies. In such case, there is nothing wrong with the solution, but rather the original problem was merely formulated indiscriminately.

SUMMARY. The formulation of a linear-programming problem has been discussed and illustrative sample problems have been presented. The simplex algorithm was introduced. The procedures for converting a linear-programming problem from the standard form, expression (1), to the simplex form, expression (2), were outlined. Possible outcomes from the simplex solution and their interpretations were presented.

The details of the simplex algorithm were not presented, but computer programs for implementing the algorithm are standard computer software. A very elementary illustration of tableaux for the simplex method was shown in Chapter 1.

Part 2. Dynamic Programming (DP)

INTRODUCTION. "Dynamic programming" is a mathematical procedure often useful for making a sequence of interrelated decisions for a system that is subject to constraints such that some function is optimized by the combination of decisions. In contrast to linear programming, there does not exist a standard mathematical formulation of the dynamic-programming problem. Rather, dynamic programming is a "family" of algorithms, wherein a general type of approach to problem solving is followed.

A degree of ingenuity and insight into the general structure of dynamic programming is required to recognize when and how it can be applied to problems. These abilities can perhap best be developed by an exposure to a wide variety of dynamic-programming applications with a subsequent elaboration on the common characteristics of the various situations and the formalations of their solutions.

A problem in dynamic programming, a transportation problem, with a simply expressed solution was given in Chapter 1. It might be advantageous co refer to that problem to get the feel of the method before continuing with the present chapter which is more mathematically oriented.

STAGECOACH PROBLEM. The "stagecoach problem" was developed for the explicit purpose of illustrating the features of dynamic programming by Prof. H. M. Wagner of Stanford University. The problem concerns a mythical salesman who had to travel west by stagecoach through hazardous territory a number of years ago. There existed a number of options as to the precise path that he could take. In order to arrive at a choice which should in all probability be safest, the salesman arrived at the clever conclusion that the costs of life insurance policies along the various links of the existing paths should be consulted. The safest route should, he reasoned, be the cheapest for life insurance coverage. The following figure shows the various paths (arrows), junction points or states (squares), and life insurance costs in dollars along each path.

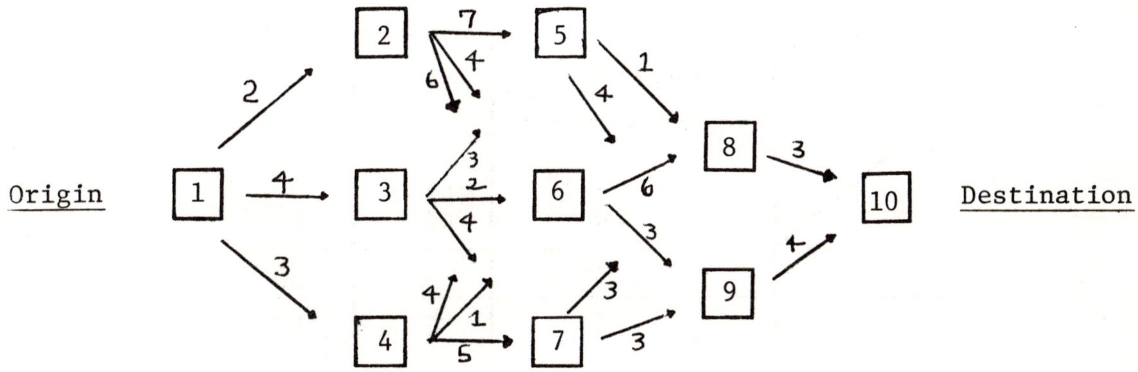

What route minimizes the total cost of the policy?

One possible approach to solving the problem is to use trial and error, an unappealing task. Dynamic programming separates a problem into stages, finds an optimal solution for the last stage, enlarges the solution to include the last two stages, etc., until all stages from the first to the last are included.

For this problem, the four stages are (1) arrival at destination 10, (2) arrival at junctions 8 or 9, (3) arrival at junctions 5, 6, or 7, and (4) arrival at junctions 2, 3, or 4. The states of a stage, denoted by "s", are the points from which the arrival originated; thus, for stage 2 the three states s = 5, 6, 7 exist, etc. The objective at each stage is to pick the best specific arrival point for each possible state; e.g., the arrival points, designated by "x_2" for stage 2, are x_2 = 8,9.

The analysis at stage 1 is completely depicted by the following "tableau" (table) in terms of the costs C_{sx1} along paths $s \rightarrow x_1$.

$s \backslash x_1$	$f_1(s,x_1) = C_{sx1}$: 10	$f_1^*(s)$	$x_1^*(a)$
8	3	3	10
9	4	4	10

$f_1(s,x_1)$ = *minimum* cost of insurance for the path from stage 2 to the final destination if the path $s \rightarrow x_1$ is a part of it.

$f_1^*(s)$ = minimum $\{f_1(s,x_1)\}$.

$x_1^*(s)$ = x_1 value at which $f_1^*(s)$ occurred.

Since only one arrival point exists at stage 1, the above tableau is quite uninteresting. Similar tableaus for stages 2, 3, and 4 follow, and they are quite more interesting. The symbols are defined similarly to those for stage 1.

$s \backslash x_2$	$f_2(s,x_2) = C_{sx2} + f_1^*(x_2)$: 8	9	$f_2^*(s)$	$x_2^*(s)$
5	4	8	4	8
6	9	7	7	9
7	6	7	6	8

	$f_3(s,x_3) = C_{sx_1} + f_2^*(x_3)$				
$s \backslash x_3$	5	6	7	$f_3^*(s)$	$x_3^*(s)$
2	11	11	12	11	5,6
3	7	9	10	7	5
4	8	8	11	8	5,6

	$f_4(s,x_4) = C_{sx_4} + f_3^*(x_4)$				
$s \backslash x_4$	2	3	4	$f_4^*(s)$	$x_4^*(s)$
1	13	11	11	11	3,4

The optimal solution can now be determined, and the minimum cost is \$11. Starting with the last tableau, the results indicate that the salesman can proceed to either junction 3 or 4. We next examine the second from last tableau; here we see that we proceed from 3 to 5 or we proceed from 4 to 5 or 6, depending on whether we initially proceeded to 3 or to 4. We proceed likewise to the first two tableaus, finding that there are the following optimal routes: $1 \to 3 \to 5 \to 8 \to 10$, $1 \to 4 \to 5 \to 8 \to 10$, $1 \to 4 \to 6 \to 9 \to 10$.

CHARACTERISTICS OF DYNAMIC PROGRAMMING PROBLEMS. The above stagecoach problem is a literal prototype of dynamic-programming problems, and therefore one way to recognize a dynamic-programming problem situation is to examine whether its basic structure is analogous to that of the stagecoach problem. The basic features are

1. The problem can be divided up into "stages," with a "policy decision" required at each stage. The stagecoach problem had four stages, each having been described and analyzed by a tableau presentation.

•2. Each stage has a number of "states" s associated with it. In general, the states are the various possible conditions in which the system might find itself at that stage of the problem.

•3. The effect of a "policy decision" at each stage is to transform the current state into a state associated with the next stage. In other words, the construction of a tableau at a certain stage, other than the first stage, is dependent on the results of previous stages.

•4. Given the current state, an optimal policy for the remaining stages is independent of the policy adopted in previous stages (i.e., subsequent tableaus). For example, if the salesman in the stagecoach problems had started at the second stage, we see from the third tableau that an optimal solution from that stage would have been the route 3→5→8→10 at a cost of $7, regardless of how he got to the second stage. This is the so-called "Markovian" property of dynamic programming.

•5. The solution procedure proceeds sequentially from the last stage of the problem to the first stage, with an optimal solution for each relevant state determined at each stage.

•6. There exist recursive relationships at each stage which determine the optimal policy from that stage forward for each state-decision variable possibility. For the stagecoach problem, the single recursive relationship for the n-th tableau (i.e., the (5-n)th stage) associated with the state s and decision variable x_n was

$$f_n(s,x_n) = C_{sx_n} + f^*_{n-1}(x_n).$$

The precise form of the recursive relationship differs quite considerably among dynamic-programming problems.

•7. An optimal solution for each state of a particular stage is determined by a relationship of the form

$$f^*_n(s) = \max_{x_n}/\min_{x_n} \{f_n(s,x_n)\}$$

sometimes with restrictions on x_n.

·8. After the optimal solution is found at the first stage (i.e., the last tableau), the overall optimal policy for the dynamic-programming problem is determined straight forwardly by sequencing back through the tableaus on a stage-by-stage basis.

PRODUCT ALLOCATION PROBLEM. The manager of a chain of three grocery stores has acquired a great deal of knowledge about the marketability (i.e., the supply-demand characteristics) of avocados at each of the stores. Having just purchased four crates of avocados, he compiled the following table of expected profit in dollars at each store, where it is assumed that crates will not be split-up:

Number	Store #		
of Crates	1	2	3
0	0	0	0
1	4	2	6
2	6	4	8
3	7	6	8
4	7	8	8

What allocation of crates of avocados will maximize the expected profit?

The problem can be treated as a three-stage problem, with the decisions at each store representing a stage. The state s of a store is the number of crates arriving at the store, and the decision variable x_n, where there are n remaining stores, represents the number of crates remaining at the n-th stage (i.e., the (4-n)-th store.) Note that the actual ordering of the stage-store relationship is immaterial; however, for simplicity, it will be assumed that the problem proceeds from store 1 to store 3, so that store 3 is considered at stage 1. The recursive decision making function is the profit associated with the optimal policy, and it is given by

$$f_n(s,x_n) = P_{4-n}(x_n) + f^*_{n-1}(s-x_n),$$

where $p_i(x)$ is the profit associated with x crates at the i-th store, and the optimal decision policy is simply

$$f^*_n(s) = \max_{x_n = 0,1,\ldots,4} \{f_n(s,x_n)\}.$$

The three-stage solution follows:

s \ x_1	$f_1(s,x_1) = P_3(x_1)$: 0	1	2	3	4	$f^*_1(s)$	$x^*_1(s)$
0	0					0	0
1	0	6				6	1
2	0	6	8			8	2
3	0	6	8	8		8	2,3
4	0	6	8	8	8	8	2,3,4

s \ x_2	$f_2(s,x_2) = P_2(x_2) + f^*_1(s-x_2)$: 0	1	2	3	4	$f^*_2(s)$	$x^*_2(s)$
0	0					0	0
1	6	2				6	0
2	8	8	4			8	0,1
3	8	10	10	6		10	1,2
4	8	10	12	12	8	12	2,3

s \ x_3	$f_3(s,x_3) = P_1(x_3) + f^*_2(s-x_3)$: 0	1	2	3	4	$f^*_3(s)$	$x^*_3(s)$
4	12	14	14	13	7	14	1,2

The optimum expected profit is \$14, and the allocation schemes for the crates to stores 1, 2, and 3 are (1,1,2),(1,2,1), (2,0,2), and (2,1,1).

PROBABILITY OF SUCCESS PROBLEM. Two new scientists are to be allocated among three research teams that are trying three different approaches to solving an urgent problem. After much hand-waving and meditation, the following probability-of-failure table was arrived at:

No. of new scientists	Team 1	2	3
0	.40	.60	.80
1	.20	.40	.50
2	.15	.20	.30

What allocation of the two scientists will minimize the probability that all three teams will fail? Note that the probability of any one of the three teams succeeding is maximized by merely allocating both scientists to team 1.

The problem can be treated as a three-stage problem, with each team represented by a stage. It is identical to the preceding problem, with the exception that the recursive decision-making function $f(s,x_n)$ is multiplicative rather than additive, and the function is to be minimized rather than maximized. Denoting the probability of failure associated with x scientists allocated to the i-th team by $P_i(x)$, the solution is:

s \ x_1	$f_1(s,x_1) = P_3(x_1)$: 0	1	2	$f_1^*(s)$	$x_1^*(s)$
0	.80			.80	0
1	.80	.50		.50	1
2	.80	.50	.30	.30	2

s \ x_2	$f_2(s,x_2) = P_2(x_2) \cdot f_1^*(s-x_2)$: 0	1	2	$f_2^*(s)$	$x_2^*(s)$
0	.48			.48	0
1	.30	.32		.30	0
2	.18	.20	.16	.16	2

s \ x_3	$f_3(s,x_3) = P_1(x_3) \cdot f_2^*(s-x_3)$: 0	1	2	$f_3^*(s)$	$x_3^*(s)$
2	.064	.060	.072	.060	1

The minimum probability of total failure is 0.06, and the optimum allocation to teams 1, 2, and 3 is (1,0,1).

INTEGER LP PROBLEM. The problem here is simply a linear-programming problem where all numbers are constrained to be integers. The problem is

Maximize: $z = x_1 + 2x_2 - x_3$

subject to the restrictions

(1) $x_1 + x_2 + x_3 = 5$

(2) $x_1 + x_3 \geq 2$

(3) x_1, x_2, x_3 = positive integers

The variables x_1, x_2, and x_3 will each be represented by a stage, with the state s representing the amount arriving at each stage and the decision variable x_n the amount allocated to the n-th variable. Here, it makes no difference whether we treat the first variable or the last variable at the first stage; for convenience, we will allocate the first stage to the first variable, etc. The optimization criteria all involve maximizing the recursive decision-making function $f_n(s,x_n)$. The solution follows:

$s \backslash x_1$	$f_1(s,x_1) = x_1$							
	0	1	2	3	4	5	$f_1^*(s)$	$x_1^*(s)$
0	0						0	0
1		1					1	1
2			2				2	2
3				3			3	3
4					4		4	4
5						5	5	5

Note that the states of any stage go only as high as 5, due to restrictions (1) and (3).

$s \backslash x_2$	$f_2(s,x_2) = 2x_2 + f_1^*(s-x_2)$					
	0	1	2	3	$f_2^*(s)$	$x_2^*(s)$
0	0				0	0
1	1	2			2	1
2	2	3	4		4	2
3	3	4	5	6	6	3
4	4	5	6	7	7	3
5	5	6	7	8	8	3

Note that decision variable of stage 2 can go no higher than 3, due to restrictions (1), (2), and (3).

$s \backslash x_3$	$f_3(s,x_3) = -x_3 + f_2^*(s-x_3)$							
	0	1	2	3	4	5	$f_3^*(s)$	$x_3^*(s)$
5	8	6	4	1	-2	-5	8	0

The solution, which could have easily been determined by inspection, is $(x_3,x_2,x_1) = (0,3,2)$, yielding $z = 8$.

DISCUSSION. There have been a number of different appearing decision-making recursive relationships exposed in the sample problems. In fact, it is possible to have different recursive relationships at each stage. However, most all dynamic-programming problems can be shown to have a basic recursive relationship for an N stage problem which is one of two forms, vis.,

$$f_n(s,x_n) = P_{N+1-n}(x_n) + f^*_{n-1}(s-x_n)$$

or

$$f_n(s,x_n) = P_{N+1-n}(x_n) \cdot f^*_{n-1}(s-x_n).$$

To facilitate the reversion of a dynamic-programming problem to one of these forms, a so-called "payout" or "cost" table such as that for the product allocation problem and the probability-of-success problem usually needs to be constructed. The term "payout" is generally used when the optimal strategy involves a maximization and the term "cost" when it involves a minimization. For the stagecoach problem the insurance-cost figure was analogous to such a table.

The integer linear programming problem reverts to the first (additive) form. The "payout" table for the problem is constructed from the relations:

$$(1)\ P_1(x_3) = -x_3$$

$$(2)\ P_2(x_2) = 2x_2$$

and $$(3)\ P_3(x_1) = x_1$$

Thus, the table for $P_n(x_n)$ becomes

x_n \ n	1	2	3
0	0	0	0
1	1	2	-1
2	2	4	-2
3	3	6	-3
4	4	$-\infty$	-4
5	5	$-\infty$	-5

Note that the $(-\infty)$ value is assigned to $x_2 = 4,5$, as x_2 must be ≤ 3, so that no possible solution can result with $x_2 > 3$.

Computer programs are available as software for solving the dynamic programming standard forms described above.

SUMMARY. Dynamic programming was introduced and exemplified with the "stagecoach" problem, and the characteristics of dynamic programming problems were enumerated with specific reference to the stagecoach problem. Three more examples of dynamic programming problems were presented, and it was shown that all of the sample problems could be "boiled down" to one of two basic forms (i.e., two basic algorithms).

Clearly, the applications of dynamic-programming techniques to relevant problems can require considerable ingenuity. Beside mathematical modeling of a particular problem, perhaps the "key" to the dynamic programming application is the recognition of the "stages" of the problem. Then the decision-making recursive expression at each stage can be formulated in terms of a "cost" or "payout" function and the optimal results of a previous stage.

The construction of a "cost" or "payout" table can simplify the dynamic programming algorithm, and it is a necessity for the utilization of a computer program. The solution procedure for dynamic programming through the usage of tableaux, presented in this chapter, is conceptually the simplest of several known procedures. However, deeper study in the field will convince one that other procedures are more efficient for more complex problems.

Generally speaking, dynamic programming is applicable to three vital areas of operations research, vis., (1) network analysis, (2) assignment or allocation problems, and (3) non-linear programming and integer linear programming. There are also applications in the important area of game theory.

Part 3. Simulation

INTRODUCTION. The technique of simulation (often called "Monte Carlo" technique) has long been an important tool of the designer, whether he be simulating airplane flights in a wind tunnel, or simulating lines of communication with an organizational chart, etc. With the advent of the high-speed digital computer with which to conduct simulated experiments, this technique has become increasingly important to the operations researcher. Simulation has been said to be the experimental arm of operations research.

The operations-research approach to problem solving always involves the formulating and solving of mathematical models which represent real systems. If the model is amenable to solution, then an analytical approach usually is superior to simulation. However, many problems are so complex that they cannot be solved analytically, and indeed, many problems are impossible to describe in mathematical terms. Thence, numerical simulation often affords the only practical approach to a problem. Furthermore, due to the many developments in digital computer technology, it is often more convenient, inexpensive, and time-saving to solve relatively simple problems by simulation.

Simulation is nothing more or less than the technique of performing sampling experiments on the mathematical model of a system. Thus, models "represent" reality and simulation "imitates" it. Simulated experiments should be viewed as virtually indistinguishable from ordinary statistical experiments, wherein stochastic processes are an intrinsic reality, and they must be based upon sound statistical theory. The heart of the statistical facet of simulation is the "random variate." An illustrative example of a trivial numerical simulation follows which should enhance the understanding of these concepts.

COIN PROBLEM. Suppose you were offered the chance to play a game whereby you would repeatedly flip an unbiased coin until a sequence of three tails or heads was tossed. You would receive $8 at the end of each game, and you would be required to pay $1 for each flip of the coin. Assuming that your supply of money is not significantly limited, should you play the game?

The statistical-analytical solution is straightforward, but tricky. Namely in any sequence of three throws, the first throw is of no significance; and, the probability of repeating the first throw two times in succesion, regardless of what the first was, would be (1/2)(1/2) = 1/4. Thus, the odds are 4 to 1 against you in any sequence of three throws. The odds are then *even* that you will win the game once in a total of two sequences. A total of two sequences costs you a *maximum* of $4, where the sequences are throws (1,2,3) and (2,3,4), an average of \$2 per sequence. However, additional sequences, when required, cost only \$1 each, as each new throw terminates a new sequence of three. Thence, the game appears to be a good deal for you, since the return on even odds is even.

A more quantitative analysis of the game can be made by simulation. One technique for simulating the game would be to play the game alone by repeatedly flipping a coin until it becomes clear whether it is worthwhile playing for stakes, a time-consuming task. A better idea would be to numerically simulate the game, using either a table of random numbers or a computer-generated sequence of ramdom numbers.

The above simulation does not necessarily confirm our suspicions about the game, although it clearly does not contradict them, because there is some finite probability that only 100 simulated flips of the coin will yield erroneous statistics. However, as the number of simulated flips increases, the probability of erroneous results becomes infinitessimal. Indeed, in reality, a computer numerical simulation would be completely automated to the extent that the above hand analysis would be programmed into the computer; and the only computer output would be the final statistics for literally thousands of simulated coin flips--all simulated in a few microseconds.

For example, assume that the computer can generate a sequence of digits 0,1,····,9, each having an equal probability of occurrence. Since the probability of a head would be one-half, we assign digits 0,···4 to the event of a head, and likewise we assign digits 5,····9 to the event of a tail. A sequence of 100 actual computer-generated digits follows: 8137271655,7900343568,5894804865,3592579729,3985892576, 9760739827,1032627137,0441832139,5965038789,5408380131. Converting to heads H and tails T and separating the games by a "/", we obtain the following numerically-simulated games: THHTHTHTTT/ TTHHH/HHTTT/TTT/HTHHTTT/HTTHTTT/THTHTTT/TTHTTT/TTT/HTHTTHTHHH/ HTHTHHHTHHH/HTHHH/HTTT/HTHHTTT/TTHHTHTHHH/H. Thus, we see that fifteen games were won, yielding $120, but the total cost was $100, yielding 6-to-5 odds of making a profit.

RANDOM VARIABLES. In the coin problem just described we made use of a computer generated series of random digits with equal probability of occurrence, a seemingly simple task.

Tables of random numbers are available. Random numbers can be generated in many pocket programmable calculators. They can, of course, be generated on command in computers. The cost engineer should understand the use of random number in simulation, and need not be an expert on the mathematics involved in their generation.

Although random numbers are readily generated with a "flat" distribution, that is equal probability of occurrence, in many problems it is necessary to generate random variates with numerous probability density functions other than flat. Computers are capable of doing this directly and can generate ramdom variates for the following distributions, among others

- ·Exponential distribution
- ·Poisson distribution

- Chi-square distribution
- Binomial distribution

QUEUEING PROBLEM. The multifarious facets and applications of numerical simulation make a detailed description of simulation techniques and practices impossible. Perhaps more insight into the workings of numerical simulation can be gained from examination of additional examples. An important field of numerical simulation is queuing theory.

Queueing theory involves the mathematical study of "queues" or waiting lines. The formation of queues is, of course, a common phenomenon which occurs whenever the current demand for a service exceeds the current capacity to provide that service. The function of queueing theory is to provide realistic mathematical models of queue systems, and often to provide analytical solutions to them. A basic queueing theory model is the so-called "single-server" model with Poisson input and exponential service time.. The following figure depicts the model:

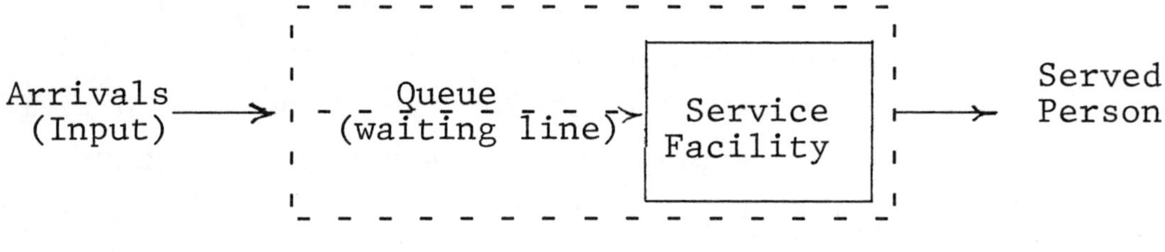

The *elapsed time between arrivals* τ has the probability density function

$$f(\tau) = \frac{e^{\frac{\tau}{\tau_m}}}{\tau_m}$$

Where τ_m is the mean τ; and the service facility requires a time t to perform the service with a probability density function

$$f(t) = \frac{e^{\frac{t}{t_m}}}{t_m}$$

where t_m connotes mean time. The problem is to determine the average waiting time in the queue. The solution ot this problem has been determined analytically and is quite well known in queueing theorty. The waiting time in the queue W_q is given by

$$W_q = \frac{\frac{t_m}{\tau_m}}{\frac{1}{t_m} - \frac{1}{\tau_m}}$$

and the probability density function of waiting time, including service is known to be Erlang.

For $t_m = 12$ and $\tau_m = 20$ it is found from the above that W_q, the expected waiting time in the queue, is 18 time units.

As an example of simulation, the problem has been computer programmed for a solution by asynchronous timing. The procedure is quite simple and is outlined as follows:

- (1) The simulation is initialized at time zero by entering a person into the queue and then immediately into the service facility, as it is initially empty. At this time the next arrival time and the service time for the initial person are determined.
- (2) Time is incremented one unit and a test is made to determine (a) if it is time for another person to enter the queue and/or (b) if it is time for another person to leave the queue and enter the service facility. Then, (a) another person is added to the queue if it is time to do so, and (b) if it is time

to do so, a person leaves the queue (if any person is waiting there) and enters the service facility. If nobody is available for the service facility, then it remains idle for at least one unit of time.

- (3) Time is again incremented one unit and procedure (2) is repeated, etc., until some prespecified number of persons has entered into the queue.
- (4) In the mean time, at each time increment an account of the number of persons in the queue is made, and the total man-time units in the queue is accumulated.
- (5) The total accumulated man-time units in the queue is then divided by the total number of persons who have been serviced to determine the average waiting time in the queue.

The results from the computer program for $t_m = 12$ and $\tau_m = 20$ are

number of arrivals	mean waiting time
1000	23.13
2000	19.39
3000	19.02
4000	18.04
5000	17.84

The simulation results are seen to range from 23.17 to 17.82 time units, generally increasing in accuracy as the length of the simulation increases.

DISCUSSION. The vast utility of simulation practices can probably be envisaged by even an amateur from the few preceding examples. The discussion has been quite general and has not treated any of the specific problems and considerations of simulation. Among these considerations are, perhaps most importantly, the steady-state condition of the simulation, the sample size, the number of samples, and sampling scheme.

The steady-state condition becomes of concern where a system being simulated operates continually in a steady-state condition. Unfortunately, a simulation model cannot be operated this way; it must be started and stopped. Because of the artificiality introduced by the abrupt beginning of operation, the performance of the simulated system does not become representative of the "real-world" system until it, too, has reached a steady-state condition. Therefore, the data obtained during this initial period of operation should be excluded from consideration. Furthermore, the simulation should be started in a state as representative of steady-state conditions as possible.

The choice of sample interval or size is quite important in planning a simulation study. For example, in the queueing problem, the sample interval was, for all practical purposes, 5 percent of the average number occurring in the simulation. For the coin problem the sampling interval was essentially infinitesimal. and imposed no restrictions on the accuracy of the results. Clearly, one's confidence in the results of the queueing problem should be enhanced by the usage of a smaller sample interval. The following table shows the results for the queueing problem where the sample interval was respectively 1 time unit and 0.1 time unit, for 20 consecutive independent simulations of 500 samples each

Run #	Sample interval (time units) 1	0.1
1	13.18	15.00*
2	20.25*	12.90
3	15.25*	11.04
4	19.36	18.85*
5	12.55	21.53*
6	12.22	13.75*
7	17.25*	20.16
8	15.93*	12.17
9	10.03	11.17*
10	12.30	15.84*

Run #	Sample interval (time units) 1	0.1
11	16.96*	14.33
12	15.38*	15.09
13	22.01	14.57*
14	11.09	17.22*
15	21.61	16.47*
16	17.02*	15.65
17	17.07*	21.08
18	9.92	14.57*
19	16.39	18.29*
20	15.74*	15.53

*Indicates closest to correct value (18.0).

The results for the smaller sample interval are closer to correct overall, and on a run-by-run basis, they were closer on 11 out of the 20 runs.

The effects of sample interval and sample length on numerical simulation results are essentially indistinguishable, especially when compensation has been made for the initial stabilization period, and a decrease in the sample interval increases the simulation calculations by the same amount as a corresponding increase in sample lenght. A so-called "confidence interval" for some specified sample interval and sample size can be established by the analysis of a sequence of simulations, which provide a sequence of statistically-independent observations. Since the observations represent averages, they should tend to have an approximately normal distribution according to the Central-Limit Theorem. Furthermore, by making the runs consecutively and using the ending condition of one run as the steady-state starting condition for the next run, the stabilization period, where it is of concern, can be limited to the single initial run. Approximately 15 to 20 observations should prove to be sufficient. The data from the above table are plotted on the following histograms:

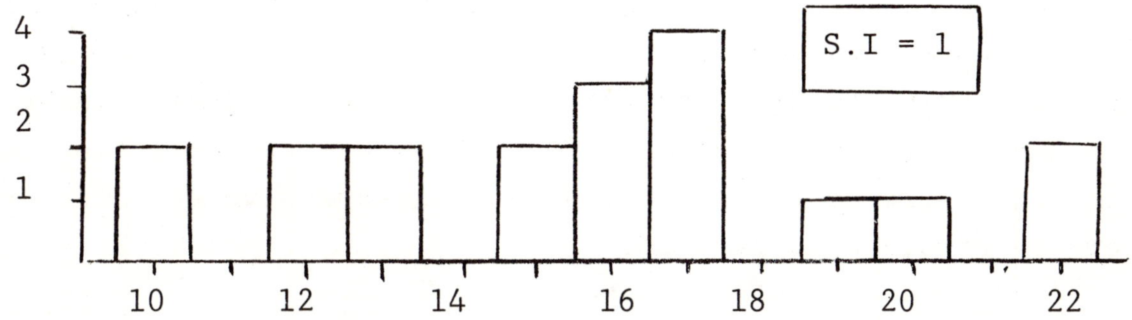

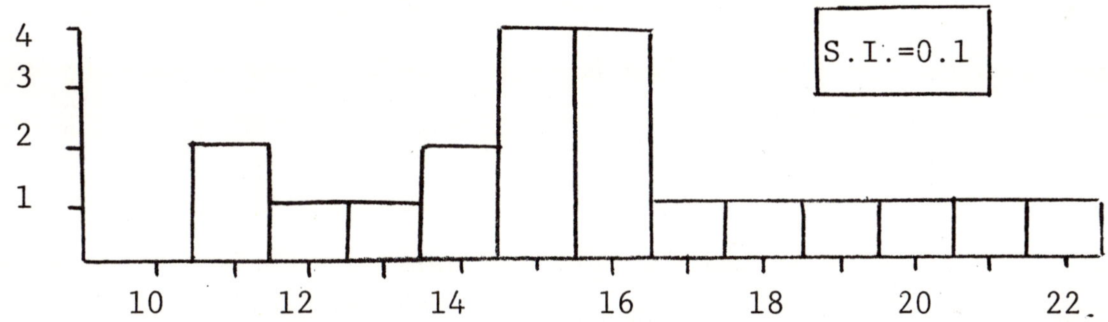

Here, we see that the variance of the observations with the smaller sample interval is somewhat smaller; also, their mean value is just slightly more correct (closer to 18.0), vis., 15.8 vs. 15.5, although the correct value would not be known for an actual simulation.

Techniques for sampling so as to decrease the variance of the above observations, resulting in more dependable results, are often called "Monte Carlo" techniques (although the term is frequently applied to simulation in general). Since these techniques tend to be rather sophisticated, it is not possible to explore them here; among these techniques are, e.g., the techniques of stratified sampling, importance sampling, and the control variate method.

In conclusion, simulation is indeed a very versatile tool, but by no means a panacea. It is invaluable for use on problems where conventional analytical techniques are out of the question, but it does not provide much insight into the cause-and-effect relationships within a system. Simulation requires a large amount of arithmetic calculations, but the new generations of computers are so fast and relatively inexpensive that this generally presents no problem.

INDEX